Development of the Hypothalamus

Colloquium
Digital Library of Life Sciences

This e-book is a copyrighted work in the Colloquium Digital Library—an innovative collection of time saving references and tools for researchers and students who want to quickly get up to speed in a new area or fundamental biomedical/life sciences topic. Each PDF e-book in the collection is an in-depth overview of a fast-moving or fundamental area of research, authored by a prominent contributor to the field. We call these e-books Lectures because they are intended for a broad, diverse audience of life scientists, in the spirit of a plenary lecture delivered by a keynote speaker or visiting professor. Individual e-books are published as contributions to a particular thematic series, each covering a different subject area and managed by its own prestigious editor, who oversees topic and author selection as well as scientific review. Readers are invited to see highlights of fields other than their own, keep up with advances in various disciplines, and refresh their understanding of core concepts in cell & molecular biology.

For the full list of published and forthcoming Lectures, please visit the Colloquium homepage: www.morganclaypool.com/page/lifesci

Access to Colloquium Digital Library is available by institutional license. Please e-mail info@morganclaypool.com for more information.

Morgan & Claypool Life Sciences is a signatory to the STM Permission Guidelines. All figures used with permission.

Colloquium Series on
The Developing Brain

Editor
Margaret M. McCarthy, Ph.D.,
Professor and Chair
Department of Pharmacology
University of Maryland School of Medicine

The goal of this series is to provide a comprehensive state-of-the-art overview of how the brain develops and those processes that affect it. Topics range from the fundamentals of axonal guidance and synaptogenesis prenatally to the influence of hormones, sex, stress, maternal care, and injury during the early postnatal period to an additional critical period at puberty. Easily accessible expert reviews combine analyses of detailed cellular mechanisms with interpretations of significance and broader impact of the topic area on the field of neuroscience and the understanding of brain and behavior.

Development of the Hypothalamus
Stuart A. Tobet, Kristy McClellan
2013

Stress and the Developing Brain
Lisa Wright and Tara Perrot
2012

Brain Development and Sexual Orientation
Jacques Balthazart
2012

Endocrine Disruptors and the Developing Brain
Andrea C. Gore and Sarah M. Dickerson
2012

Development of the Hypothalamus

Stuart A. Tobet, Kristy McClellan

www.morganclaypool.com

ISBN: 9781615045129 print
ISBN: 9781615045136 ebook

DOI 10.4199/C00079ED1V01Y201303DBR010

A Publication in the Morgan & Claypool Publishers series
COLLOQUIUM SERIES ON THE DEVELOPING BRAIN
Lecture #10
Series Editor: Margaret M. McCarthy, University of Maryland School of Medicine

Series ISSN 2159-5194 Print   2159-5208 Electronic

Development of the Hypothalamus

Stuart A. Tobet
Colorado State University

Kristy McClellan
Buena Vista University

COLLOQUIUM SERIES ON THE DEVELOPING BRAIN #10

 MORGAN & CLAYPOOL LIFE SCIENCES

ABSTRACT

The involvement of key factors operating independently or in cooperation with others contributes to physical and physiological mechanisms to help engineer a vertebrate hypothalamus. The actions of these key factors influence developmental mechanisms including neurogenesis, cell migration, cell differentiation, cell death, axon guidance, and synaptogenesis. On a molecular level, there are several ways to categorize the actions of factors that drive brain development. These range from the actions of transcription factors in cell nuclei that regulate the expression of developmental genes, to external factors in the cellular environment that mediate interactions and cell placements, and to effector molecules that contribute to signaling from one cell to another. Sexual dimorphism is a hallmark of the vertebrate hypothalamus that may arise as a direct consequence of hormone actions or gene actions. These actions may work through any of the mechanisms outlined above. Given the arrangement of cells in groups within the hypothalamus, cell migration may be one particularly important target for early molecular actions that help build the bases for appropriate functions.

KEYWORDS

hypothalamus, steroid hormones, cell migration

Contents

Acknowledgments

The authors would like to thank our co-workers over the years that have provided a wealth of data and discussions without which this review would not be possible. In addition, we would like to thank Drs. Philip Schlup and Steve Benoit for providing the animations of cell behavior in response to cues. Finally, we thank funding agencies over the years for providing for aspects of research that are in part reviewed as part of this discussion (e.g., MH57748 (collaboration with Dr. Keith Parker), MH082679 (collaboration with Drs. Handa and Goldstein), TW005922 and MH61376 (collaborations with Dr. Gregor Majdic), DC009034 (collaboration with Dr. Gary Schwarting), NSF 0534608 (collaboration with Dr. Colin Clay), and the Keck Foundation (collaboration with Dr. Randy Bartels)).

CHAPTER 1

Introduction

So perhaps you want to build a hypothalamus . . .

Many papers have described the development of the hypothalamus from a variety of perspectives. These have utilized multiple techniques and numerous species across vertebrate phylogeny. The hypothalamus has been subdivided by different investigators into compartments, units, cell groups, or nuclei that range in number from three (compartments) to almost two dozen (cell groups).

Consider the characterization that the brain has evolved to use groups of cells to carry out three fundamental roles that can be traced to single cells: sensory input, information integration, and motor output (Nauta and Feirtag, 1979). In the brain as a whole, sensory functions start with primary sensory neurons (frequently, but not always, located in peripheral ganglia) and motor functions within specific groups of motor neurons (primarily in the brainstem and spinal cord). Integrative functions are assigned to what WJH Nauta called the "great intermediate net," which in mammals comprises most of the brain. Interestingly, the hypothalamus is one place where the three major components of brain function are represented together. Thus, the hypothalamus contains direct sensory capacity (e.g., for light, glucose, and blood osmolarity among others), tremendous "intermediate net" processing capacity (e.g., neuroendocrine or autonomic integration), and direct motor capacity (e.g., secretory neurons).

In general, the hypothalamus is considered a key brain site for the regulation of numerous homeostatic functions (Lechan and Toni, 2008). Many researchers have indicated clinical significance for hypothalamic dysfunction both developmentally and in adulthood (Schindler et al., 2012; Swaab 2004; Caqueret et al., 2005; Michaud, 2001). Interestingly, some evidence suggests that developmental analyses of human hypothalamic anatomy may provide better comparisons with other species (Koutcherov et al., 2002; 2003). The study of human hypothalamic development from 9 weeks gestation through birth examined the timing of development through examining the presence of various cell types. The results of these human studies provided evidence for close homologies with the rodent hypothalamus in both timing and neurochemistry.

This book explores the complex and organized processes that go into building a hypothalamus. The processes of development will be highlighted with a focus on cell migration as a mechanism for the development of cell groups.

CHAPTER 2

Compartments in the Diencephalon

Early nervous system development requires changes in cell behavior (gene expression, cell-cell interactions, migration) to divide the neural tube into multiple compartments, including hindbrain, midbrain, and forebrain. Early neural tube development occurs following the period of gastrulation when the first three tissue layers of the embryo are being established. The organization of the neural tube emerges soon after closure with dorsal/ventral and anterior/posterior polarity. Over time, lateral expansion of the neural tube gives rise to the forebrain, which emerges from the rostral portion of the neural tube. The hypothalamus exists as a compartment of the diencephalon. As development progresses over time, the telencephalon becomes the cerebrum while the diencephalon subdivides into the thalamus, hypothalamus and epithalamus (Puelles and Rubenstein, 1993). Growth of the neural compartments occurs differentially, leading to a larger forebrain region in comparison to the midbrain and hindbrain. Figure 2.1 shows the difference in size and number of cells occupying the region of the hypothalamus between embryonic day 15 (E15) and postnatal day 0 (P0).

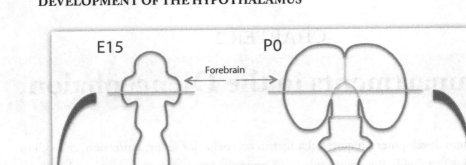

Figure 2.1: The neural tube develops into the central nervous system with the forebrain subdividing into the telencephalon and diencephalon. Top row—Dorsal view of developing mouse central nervous system. The forebrain is labeled and is the most anterior portion of the neural tube. The spinal cord is the furthest posterior. Bottow row—sample of coronal section found within the forebrain (location of section is indicated by a line). Region of the hypothalamus is found at the base of the coronal sections and is highlighted by the orange box. The hypothalamus, a subdivision of the diencephalon, begins to form distinct nuclear groups as early as embryonic day 15 (E15) in the mouse.

2.1   SPATIAL PATTERNING OF THE HYPOTHALAMUS

In adults, subdivisions of the hypothalamus form functional groups, cells that organize with defined boundaries, and related cell identities controlling related physiology and behaviors. During development, these subdivisions may also indicate groups with similar cell histories (Puelles et al., 1987; Puelles and Rubenstein, 1993). Telencephalic cells are thought to originate from progenitors in the proliferative zones adjacent to the lateral ventricles and diencephalic cells from progenitors in proliferative zones surrounding the third ventricle. Experiments describing the distribution of transcription regulating genes (reviewed Szarek et al., 2010) or the migration of dye-injected cells in embryonic chicken brains (e.g., Figdor and Stern, 1993) suggest that such developmental subdivisions may constitute physical compartments with boundaries. Therefore, cells born in one compartment may be restricted from entering neighboring compartments (Fishell et al., 1993).

On the other hand, there are exceptions and data showing the pattern of radial glial fibers indicate the potential for extensive cross-compartmental mixing in the rostral hypothalamus (Tobet et al., 1995). More recent data shows extensive contributions of cells originating in the preoptic area that lies at the rostral end of the hypothalamus to cortical and hippocampal locations (Jaglin et al., 2012; Gelman et al., 2011)

The anatomical organization of the diencephalon is visible through Nissl stains as individual groups (or nuclei) of cells surrounded by cell-poor zones. These nuclear groups have also been described based on their cellular chemistry, connections, and regulation of specific functions (Canteras et al., 1994; Segal et al., 2005). The diversity of cell phenotypes within the nuclear groups contributes to the regulation of functions and behaviors associated with the hypothalamus (Figure 2.2). For example, the ventromedial nucleus of the hypothalamus (VMN) is characterized by the expression of the transcription factor steroidogenic factor-1 (Parker et al., 2002). The paraventricular nucleus (PVN) also has an associated transcription factor that is somewhat selective though not nearly as specific for PVN as SF-1 is for the VMN (Michaud et al., 1998). The PVN also contains numerous characteristic peptides, including vasopressin (anti-diuretic hormone) and oxytocin synthesizing cells (Armstrong et al., 1980), while the supraoptic nucleus (SON) contains a second group of magnocellular secretory cells synthesizing vasopressin and oxytocin. The arcuate nucleus (ARC) is known for orexigenic and anorexigenic peptides controlling food regulation (Bouret and Simerly, 2004). All of these nuclear groups are considered part of the circuitry that is critical for feeding and energy balance.

The suprachiasmatic nucleus (SCN) has an unusual group of cells that receive input from the retina and are responsible for generating circadian rhythms. Circadian rhythms govern multiple homeostatic functions including physiology and behavior that ranges from hormone release and body temperature changes to sleep, wake, and activity rhythms.

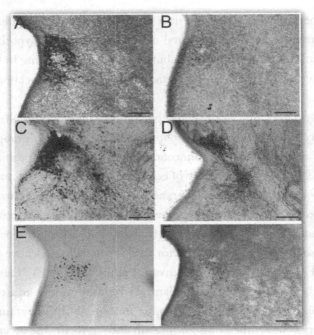

Figure 2.2: Nuclear groups within the adult hypothalamus have the distinct appearance of boundaries using Nissl stains and are comprised of groups of cells performing related functions. In embryogenesis, distinct nuclear groups are not clearly defined by Nissl stained defined boundaries, even though functional cell groups based on cell phenotypes are discernible. The PVN is associated with numerous cell phenotypes as early as Embryonic Day 15 (E15). Corticotropin releasing hormone (A), vasopressin (B), calbindin (C), neuronal nitric oxide synthase (D), estrogen receptor β (E), estrogen receptor α (F), galanin (G), and neuropeptide Y (H) demarcate distinct populations of cells in the PVN at early ages (scale bars = 50 um). (adapted from McClellan et al., 2010)

Neurons in the preoptic area and anterior hypothalamus (POA/AH) participate in regulating many homeostatic and sex-dependent functions in adults. It is a region that contains numerous sex differences in neural morphology (reviewed by Tobet and Fox, 1992). The POA/AH sits at a presumptive border of the diencephalon and telencephalon; thus, there has been some ambiguity about its developmental origin (Creps, 1974; Swanson and Cowan, 1979). The region begins rostrally and ventrally with the lamina terminalis and merges caudally with the medial anterior hypothalamic area above the caudal portion of the optic chiasm, and the medial division of the bed nucleus of the stria terminalis (BNST). The anterior commissure is normally taken as its dorsal border. Neuronal birth-dating studies using [3H] thymidine in rats have indicated that cells in the medial preoptic area are derived primarily from progenitor cells in proliferative zones around the third ventricle (Altman and Bayer, 1978; 1986; Bayer and Altman, 1987). By contrast, similar studies suggest that cells in the adjacent BNST may be derived from precursors in proliferative zones surrounding the lateral ventricles (Bayer, 1979, 1987).

2.2 TEMPORAL PATTERNING OF THE HYPOTHALAMUS

Much of the data regarding the timeline for development of the hypothalamus is found in studies using the mouse or rat. Neurons that populate the hypothalamus are derived from progenitor cells of the subventricular zone between E9 and E17 in mice, E11–E19 in rats (gestation being about 19-21 days). In general, cells that populate the nuclei of the hypothalamus are born along the subventricular zone and migrate to populate the individual nuclear groups. Birthdating studies (following cells based on retention of identifiable nucleic acid bases after their final mitotic division; Miller and Nowakowski, 1988) looking at each time-point during embryogenesis have been useful in determining when individual nuclei emerge. Neurons that populate the VMN are born between E10 and E15 in mice, E13–E17 in rats, and around E30 in the primate (Shimada and Nakamura, 1973; Tran et al., 2003). The more dorso-ventrally located PVN is comprised of a magnocellular (lateral) group of cells that are primarily born between E10.5 and E12.5 in the rat while the parvocellular (medial) group of cells are born later (Altman and Bayer, 1986). This brings up one of the features of the organization of the diencephalon. In general, cells that are located in more lateral positions within the hypothalamus originate from the subventricular zone before those cells located in more medial positions (Markakis & Swanson, 1997). This separates the organization of the diencephalon from that of telencephalon, in which the cerebral cortex develops as a layered structure with early born cells located closest to the ventricular zone.

2.3 GENES INVOLVED IN EARLY HYPOTHALAMUS SPECIFICATION

At each stage of development, the genes expressed within a neural region represent the state of its molecular specification. These gene expression patterns characterize the regional organization of the brain and are involved in the regulation of the main developmental processes: neuronal proliferation, death, migration, differentiation, and establishment of connections. The result is the establishment of anatomical regions or nuclear groups that are associated with specific physiology or behaviors.

One fundamental question in studying the development of a particular structure is determining which factors (genes, transcription factors) are involved in specifying the region. Examining the expression patterns of known developmental markers gives us some idea of potential regulators of hypothalamic development. A recent large-scale study examined gene expression over 12 time points within the developing hypothalamus. The results were used to create a molecular atlas of the developing hypothalamus highlighting spatial and temporal changes in a large a number of genes (Shimogori et al., 2010). The results of this study provide a resource for further investigation of hypothalamic development. The identification of the molecular players in hypothalamic development will lead to a better understanding of the processes involved; migration, differentiation, and connectivity.

Experiments utilizing various knockout or transgenic mice have been used to demonstrate the significance of particular developmental genes. Otp is a homeobox gene with an expression pattern that is restricted to the developing hypothalamus. A loss of Otp expression in mice results in a loss of expression of various cell phenotypes including oxytocin, vasopressin, and dopaminergic cells of the PVN. Like many factors important in development, Otp may be involved in at least two developmental processes: terminal differentiation of neurons and migration from the ventricular zone (Acampora et al., 2000). A second gene involved in early patterning of the hypothalamus, Lhx2, is part of the LIM family of transcription factors and is expressed within areas of the anterior hypothalamus of mice and zebrafish. The targeted deletion of Lhx2 in mice leads to the loss of structures of the ventral hypothalamus, including the absence of the median eminence. This gene provides a good example of the cascading nature of gene knockout effects. A recent report shows the loss of neurons containing the peptide gonadotropin-releasing hormone (GnRH) in mice for which Lhx2 was selectively disrupted in olfactory sensory neurons (Berghard et al., 2012). The loss of GnRH neurons was secondary to the loss of their migratory pathway with disruption of the olfactory bulb.

Early expression of a number of growth factor genes has been shown to have a role throughout the developmental process. Fibroblast growth factor 8 (FGF8) is a good example. FGF8 is a member of a large family of growth factors that signal through membrane tyrosine kinase receptors, and have important roles in early forebrain development. Suppression of FGF8 signaling in the forebrain results in indistinct boundaries of the telencephalic and diencephalic junctions. This suppression was mediated through the loss of other developmental factors (sonic hedgehog and Gli3; Rash et al., 2011). Interestingly, FGF8 continues to be important throughout development of the hypothalamus, demonstrated through temporal regulation of FGF8 expression or loss of function mutations. FGF8 is important in the development of the GnRH system, specifically in determining the differentiation and fate of GnRH neurons (Chung and Tsai, 2010). FGF8 deficient mice also exhibit a loss of oxytocin and vasopressin within nuclei of the developing murine hypothalamus, indicating a potential role in the establishment of the PVN and SON. Medical case studies illustrate the importance of FGF8 in human development. Loss of FGF8 expression can result in developmental abnormalities including holoprosencephaly, craniofacial defects, and hypothalamo-pituitary dysfunction (McCabe et al., 2011).

CHAPTER 3

Building a Hypothalamus: Neurogenesis, Cell Survival, and Cell Death

There are several components necessary to build any brain region, hypothalamus included. Specific details change, but key points are repeated. First and foremost, the cells needed to populate the developing hypothalamus need to be produced. For this, we need to consider the process of neurogenesis (Altman & Bayer, 1986). In the 21st century, it is impossible to consider neurogenesis without considering a role for "glial" cells. For new cells to be useful, they must be moved into appropriate positions to participate in hypothalamic cell groups and circuitry. Not all cells that are born are destined to become part of the adult hypothalamus. Just as in other brain regions, many cells undergo processes of programmed cell death as part of natural development (Forger, 2009; Tsukahara, 2009). Cell migration is a key process to move cells from sites of final mitotic division to sites where they can carry out differentiated functions (Tobet et al., 2009). The cells of the hypothalamus are rich in their heterogeneity of cell phenotypes (Markakis, 2002). This is particularly evident at the level of peptide content as some cells might make vasopressin (AVP), while others make oxytocin (OXY) or corticotrophin releasing hormone (CRH). It is further amplified by the content of particular transcription factors (Szarek et al., 2010) that might include steroid hormone receptors (e.g., estrogen, androgen or progestin receptors) or others (e.g., steroidogenic factor 1 (SF-1) or COUP-TF1). Finally, as for any brain region, it is critical to develop the appropriate connections both into and out of specific hypothalamic nuclear groups (Hutton et al., 1998; Polston & Simerly, 2006).

3.1 NEUROGENESIS: DEVELOPMENTAL

Most new neurons in the central nervous system are generated close to cerebral ventricles. Populations of progenitor cells reside in proliferative ventricular zones, or subventricular zones adjacent to the ventricles. The subventricular zone is considered a secondary zone of neurogenesis, which forms in the later stages of embryogenesis and is present in all mammals. This proliferative region is responsible for the generation of a majority of glial cells and some neurons of the forebrain. It also plays a role in post-embryonic neurogenesis, which is discussed below. As noted above, most of the cells in the hypothalamus are derived from progenitor cells surrounding the third ventricle of the diencephalon. However, there are two notable exceptions that will be discussed relative to migration later in this review. These are cells that are generated either in a telencephalic proliferative zone dorsal to the anterior commissure (Tobet et al., 1995) or phenotypically identified neurons that synthesize the decapeptide gonadotropin-releasing hormone (GnRH) that arise from progenitor cells in the nasal compartment and migrate long distances caudally to reach the hypothalamus (Schwanzel-Fukuda & Pfaff, 1989; Wray et al., 1989). In humans, this migration is particularly long, and virtually all the GnRH neurons reach back further into the hypothalamus than in rodents and other species (e.g., Kim et al., 1999).

3.1.1   GROWTH FACTORS: BRAIN-DERIVED NEUROTROPHIC FACTOR

In conjunction with neurogenesis, cell survival is an important part of the development of any structure. Growth factors, such as brain-derived growth factor (BDNF) are found early in development within the region of the hypothalamus. BDNF is a member of the neurotrophin family of proteins and is a key neurotrophic factor controlling diverse brain functions in development and adulthood. The expression pattern of BDNF protein follows the embryonic timeline for hypothalamic development peaking around birth in the rat and rapidly declining after the first week (Sugiyama et al., 2003). In cortical and cerebellar development, BDNF demonstrates potential chemokinetic and chemotactic roles on migratory neurons (Behar et al., 1997). There is no direct evidence that BDNF plays a similar role influencing migratory neurons within the hypothalamus. Instead, BDNF likely influences hypothalamic development through trophic actions on neurons. The temporal expression patterns of BDNF within the VMH and PVN also suggest a role in cell differentiation (McClellan et al., 2008, 2010; Carbone & Handa, 2012). Interestingly, in adult rats BDNF administration may increase the number of new neurons found within the hypothalamus (Pencea et al., 2001).

3.1.2   NEURON VS. GLIAL PRECURSORS

The answer to the question of the identity of neuronal precursors in the developing brain has been evolving over the last decade (Ever & Gaiano, 2005; Rakic, 2006). Radial glial cells have a clearly defined structure and long history of study that stretches back to Ramon y Cajal (Puelles, 2009). They have cell bodies in ventricular zones and fibers that stretch to end feet at pial surfaces giving the appearance of a wheel with a hub (VZ) and spokes (radial cells) structurally supporting the early hypothalamus. Elegant electron microscopic reconstructions in the early 70's suggested cortical neurons migrating in direct contact with these fibers (Rakic, 1972). Until, however, the advent of studies in the beginning of this century using molecular genetics and live video microscopy the prevailing thought was that radial glia were glia and they gave rise to a separate lineage from neurons (e.g., Levitt et al., 1980). It is now clear that for the cerebral cortex, radial glia are neuronal precursors.

Radial cells are heterogeneous being immunoreactive for a monoclonal antibody called RC2 (originally named as "radial cell antibody 2" (Misson et al., 1988) that potentially recognizes a product of the nestin gene (Park et al., 2009) and has been considered a "definitive" radial cell marker in mice), differing expression of GLAST (astrocytes specific glutamate transporter), and BLBP (brain lipid binding protein). Only the subsets containing all three markers are considered radial glial cells. Radial cells can be shown to divide based on immunoreactive BLBP to show extension to the pial surface and bromodeoxyuridine (BrdU) incorporation or cdc2 kinase phosphorylated vimentin expression to show active cell cycling. The percentage of DiI labeled (sorted), RC2+ cells that incorporate BrdU do not change following 1, 3, or 5 h indicating that radial cells that divide do not let go of the pial surface (Hartfuss et al., 2001). GFP-viral infections in vivo combined with video microscopy in vitro also showed that radial cells divide and give rise to neurons that migrate along their processes (Noctor et al., 2001). One way to reconcile the asymmetry of a division that can give rise to a neuron and a progenitor/glial cells is to involve a protein known to impact the determination of cell fates in a number of systems: e.g., Notch. It is thought that notch and delta are distributed to radial cells as part of an asymmetric division. Notch action may be critical for interactions between migrating cells and the radial processes along which they migrate. Notch expression in radial cells and delta expression in migrating cells likely promotes the retention of radial cells in a progenitor state (Gaiano et al., 2000).

Just how broad is the characterization of radial cells as neuronal precursor? Radial glial cells have been noted throughout the brain including the hypothalamus for a number of years (Levitt & Rakic, 1980). To this point, and despite one broadly titled report (Anthony et al., 2004), there is little direct evidence demonstrating that radial glia are neuronal precursors in the hypothalamus. Nonetheless, this is the likely current default concept. There is another twist on radial glia-like cells in the hypothalamus in the form of tanycytes (Flament-Durant & Brion, 1985). These cells, which in the hypothalamus are ependymal cells (without cilia) lining the ventral half of the third ventricle, are morphologically similar to radial glia except for their cell bodies and persistence into adulthood. In adults, these cells have been suggested to play roles in regulating nerve terminal access to the capillaries of the median eminence (King & Rubin, 1994; Rodriguez et al., 2005; Prevot et al., 2010). Recent data suggests that tanycytes might provide a neurogenic niche in the adult hypothalamus (Xu et al., 2005; Lee et al., 2012; Saaltink et al., 2012).

3.1.3   SEX DEPENDENCE

Early findings in rats indicated that there were more early born cells destined for the sexually dimorphic nucleus of the rat preoptic area in females than males (Jacobson & Gorski, 1981). There were no sex differences in migration for later born cells (Jacobson et al., 1985) and this was recently confirmed (Orikasa et al., 2010). Experiments with mice have indicated further that there may be hormone-independent sexual differentiation of newborn cells by examining an index of age-dependent cell proliferation. Pregnant mice were administered BrdU on E11, E12, or E13 and the number and position of labeled cells on P0 was examined. For those cells born on E11 that were still present on P0 in the preoptic area, more cells were present in females than males whether the gonad was present or not (Knoll et al., 2007). Because the sex difference in this study was present regardless of gonadal status, it suggests that the presence of more early born cells in females (mice) is a hormone-independent characteristic of preoptic area development (Majdic & Tobet, 2011). More recently, 5-ethynyl-2'-deoxyuridine (EdU) became available as a substitute for BrdU. EdU utilizes Click chemistry that does not require the use of DNAse or harsh acid treatment for visualization and is more amenable for double labeling (Salic & Mitchison, 2008). In a preliminary experiment, EdU administered at E10 revealed more medial cells labeled in the preoptic area of male versus female when visualized at E17 (see Figure 3.1). The pattern is not identical to the earlier result with E11 BrdU examined at P0 (Knoll et al., 2007), and further studies are needed to clarify the differences.

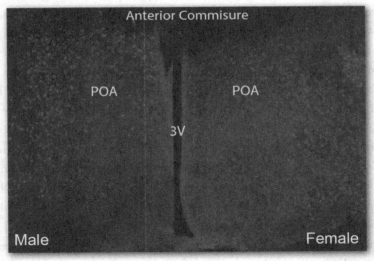

Figure 3.1: EdU administered at E10 revealed more medial cells labeled in the preoptic area of males vs. females when visualized at E17 (figure is from the work of Stratton and Tobet).

3.2   CELL DEATH

Programmed cell death plays a major role in the formation of many brain structures (Buss et al., 2006). In the mid 1990s it was first reported that there might be significantly more cell death during the normal development of cerebral cortex than previously appreciated (Blaschke et al., 1996). Subsequent experiments revealed cell death in regions of the hypothalamus that are sexually dimorphic (Chung et al., 2000; Waters & Simerly, 2009), although not on par with the extensive levels in the cerebral cortex. Sex differences in cell death have emerged as a central component of many theories of brain sexual differentiation (Forger, 2009; Tsukuhara, 2009). There is contrasting information on cell death in the different regions of the hypothalamus that may be due in part to specific strain differences. In Sprague–Dawley rats, there was little apoptosis during postnatal development. As the ventromedial nucleus increased in size from PN2 to PN12, the amount of cell death decreased (Chung et al., 2000). However, when cell death was examined in the ventromedial nucleus of Wistar rats at postnatal ages there was a greater incidence of cell death observed in these rats at birth, and again a decrease in the number of apoptotic cells at later postnatal ages. In the preoptic area, species comparisons also lead away from a total generalization that sex differences in cell death always contribute to sexual dimorphism (Park et al., 1998).

Based on the locations of dying cells, there does not appear to be a major role for cell death in the emergence of the morphology of specific hypothalamic nuclei as there is little indication of more cell death at the periphery of forming nuclear groups than in any other particular region of individual nuclei. This is even true for the sexually dimorphic nucleus of the preoptic area where the locations of dying cells are not topographically related to sculpting the final nuclear area. On the other hand, a recent study suggests that excess early exposure to glucocorticoids (e.g., using the synthetic glucocorticoid dexamethasone) may induce cell death in the region surrounding rather than within the PVN in rats at birth (Zuloaga et al., 2012). In our own examination of cell death in and around the ventromedial nucleus, there was no obvious connection in location relative to the emergence of that cell group. Pyknotic cells (an indicator of apoptosis) were counted in VMN sections and no more than four cells were found per each 50μm thick section (Davis et al., 2004). Caspase-3 is a widely used marker of apoptotic cells and in the region of the VMN at P0; few cells express this marker in wild-type mice during early development (Figure 3.2), again suggesting only a minor role for programmed cell death in the formation of hypothalamic cell groups.

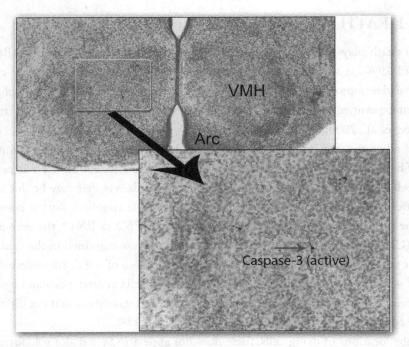

Figure 3.2: A coronal section through the hypothalamus at postnatal day 0 in a control C57BL/6 mouse was processed for immunoreactive activated caspase 3. The blue arrow indicates an activated caspase-3 immunoreactive cell at high magnification. Few activated caspase-3 immunoreactive cells are in the region of the VMH (adapted from McClellan et. al., 2006).

CHAPTER 4

Building a Hypothalamus: Cell Migration

4.1 OVERVIEW

The movement of cells from one location to another is fundamentally a question of cytoskeletal dynamics and signals that drive cells to change their cytoskeletal dynamics (Heng et al., 2010). This can be envisioned as factors that cause polymerization or depolymerization of either actin or tubulin into actin filaments or microtubules, respectively. For example, when a signal causes polymerization of actin filaments it will also cause growth or movement toward higher concentrations of the signal such as in this series of videos (Figure 4.1; Supplemental videos 4.1 and 4.2). As the actin polymerizes, it gives the potential for movement to the cell. Now that a cell has the capacity to move or migrate, decisions must be made regarding direction, speed, and placement all within the context of its surroundings and nearby cells. Factors that influence the probability of motion may work differently than those that influence the rate of motion. Figure 4.2 illustrates a model for factors affecting rate vs. probability of motion in GnRH neurons of the hypothalamus. This figure is representative of how neurons initiate movement along a surface. Factors necessary for movement work at the site of adhesion of a cell that is not currently moving. Initiation of movement requires breaking the adherence of the neuron and rearrangement of the cytoskeleton to affect the rate of movement.

Video 4.1: This video depicts the consequences of cell exposure to both an attractive cue (blue) and a "repellent" cue (white) at the same time. Yellow indicates actin polymerization. As time moves along, the repellent cue prevents actin polymerization, while the attract cue promotes actin polymerization. The result is movement in the direction of increasing polymerization (video provided by Dr. Steve Benoit).

Video 4.2: This video depicts the activation of a cell (center) that then secretes an attractive factor (red) that causes activation of a receptor on surrounding cells and causes them to move closer to the original activated cell (video provided by Dr. Philip Schlup).

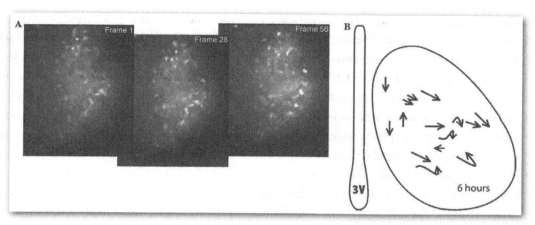

Figure 4.1: Live video microscopy of an E14 slice from a mouse expressing YFP in Thy-1 positive cells (A) was used to assess rate and direction of migrating neurons. Three representative frames show the labeling of individual neurons in the region of the developing VMH. (B) A schematic diagram shows the movements of 14 individual cells in 3 different video sequences over the course of 6 hours. As illustrated in the diagram, some cells appear to be moving dorsal/ventral while others are moving in the direction of the radial glial fibers (medial to lateral at an oblique angle). It is common for the cells to turn during the migration process (adapted from McClellan et al., 2006).

FigureKey aspects of neuronal migration include cell movement as defined by chemokinesis (changes in speed) and chemotropism (changes in response to a chemical or cue). Guidance can be considered in relation to glial cells or fibers, neuronal fibers, or even just the extracellular matrix. Molecules related to movement fall into broad categories. These include cell autonomous factors such as cytoskeletal elements that influence actin or tubulin polymerization and including all of the cytoskeletal associated proteins. They include factors external to cells in extracellular matrix (e.g., collagens, proteoglycans, or glycoconjugate adhesion molecules). They further include molecules that mediate contact mediated interactions between cells and matrix or other cells such as those important for cell adhesion (e.g., membrane glycoproteins that include superfamilies like immunoglobulins or cadherins). Finally, they include signal transduction molecules from the ligand side (e.g., GABA) and the receptor side (e.g., linkage via membrane receptors to cytoskeleton and other kinase or calcium-based signal transduction pathways).

Cell migration can be divided into several steps. First, the leading edge of a cell "contemplating" movement extends a process. This happens in tandem with adhesion of the cell to a matrix or support that could be the process of another neuron or a radial fiber, which might have a partly glial nature. The cell about to move will contract the contents of its cytoplasm (polymerize more actin in its cytoskeleton) towards the leading edge and then release from the adhesive contact sites on the opposite end. Last, there is recycling of membrane receptors from the rear to the front of the cell. This process of attachment and reattachment occurs many times along the migratory route of each cell. Factors influencing the movement of cells are likely targeting the arrangement of the cyto-skeleton (see supplemental video 1), the attachment of a cell to it's contact surface, or the recycling of membrane receptors which are needed for a second round of cell movement. In some instances, this dynamic and organized process is disrupted causing migratory disorders. Obvious migratory disorders such as Lissencephaly, which results in the appearance of a "smooth brain," can cause major cognitive and behavioral changes. Severe cases of abnormal cell migration within the nervous system are rare, yet these point out the importance of examining cell migration in relationship to the importance of cell positioning within the brain. Questions dealing with minor changes in cell migration are more difficult to address. A major migratory disruption would likely result in death as the functions of the hypothalamus are critical to maintaining energy homeostasis and autonomic function. It is intriguing to consider that minor changes in proper migration may lead to changes in structure and function within nuclear groups (Aujla et al., 2011). We have discussed above how changes in cell phenotype can lead to alterations in energy regulation. Cell migration can precede or intertwine with the process of fate determination. There are many examples in development of the importance of extracellular factors from neighboring cells in the process of cell differentiation. A cell's position within a nuclear group may be critical to connectivity and ultimately function of the cell group. A loss of the nuclear transcription factor SF-1 leads to rearrangement of cells within the VMN (Dellovade et al., 2000) and ultimately leads to a change in weight regulation and energy balance (Majdic et al., 2002). Relative to the connectivity issue, there are many fibers in the region of the VMN that are altered in their trajectories in the absence of SF-1 and the normal organiza-tion of the cells in the region (Budefeld et al., 2011).

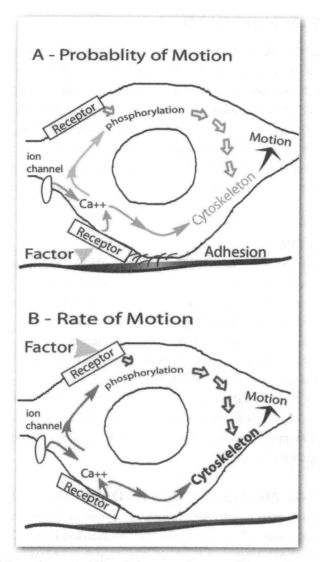

Figure 4.2: The probability of motion (A) and the rate of motion (B) are compared. In (A) factors necessary for movement work at the site of adhesion of a cell that is not currently moving. Initiation of movement requires breaking the adherence of the neuron to another surface while influencing the rate of motion (B) involves rearrangement of the cytoskeleton. As factors influence rate they act on the cytoskeleton to promote assembly in a specific direction (taken from Tobet and Schwarting, 2006).

4.2   VISUALIZING MIGRATION

The ability to visualize the movement of cells away from the proliferative zone is critical to understanding how the brain forms a complex series of layers and cell groupings. Methods utilized to visualize migration have changed as new technologies and compounds are introduced. Early studies have been aimed at determining birthdates for dividing cells in the hypothalamus. More recently, microscopy techniques have been developed to visualize migration and assess the effect of various chemical factors on speed, direction, and likelihood of movement. For the hypothalamus, these studies for the most part have been restricted to ex vivo slice cultures in mammals. For brain regions closer to the surface (e.g., cerebral cortex), more recent studies have been able to visualize cell in vivo.

4.2.1   BIRTHDATING

Neuronal birthdating is a way to track neuron migration. This is based on the short half-life of analogs used and the ability to dilute out label with multiple cell divisions. Birthdating studies were used to determine a point of time when post-mitotic cells of the subventricular zone begin to move out and populate areas of the hypothalamus. Dual tracer (3H-thymidine + BrdU) birthdating allows tremendous timing accuracy and provided a way to determine more specifically the contribution of cells born on different cycle numbers (Caviness et al., 2003). Dual tracing has also been used to examine neurogenesis in the adult mouse. The use of a dual labeling protocol using two different analogs allows tracking of proliferating cells and assessing cell cycle re-entry of neural stem cells at different time points (Moreno-Estellés M et al., 2012). The relatively newer 5-ethynyl-2'-deoxyuridine (EdU) is now more readily available and more amenable for double labeling (Salic & Mitchison, 2008; Zeng et al., 2010).

4.2.2   TRANSGENIC MOUSE MODELS AND VIDEO MICROSCOPY

The migration of neurons from proliferative ventricular zones to the region of the hypothalamus requires cues to help determine direction, speed, and boundaries. Expression of certain genes in development is important in setting up boundaries for migrating neurons (Rubenstein and Rakic, 1999). Methods involving transgenic mice with cells fluorescently labeled using promoter-driven fluorescent proteins have been created that may help elucidate these cues and their influences on migrating cells (Bless et al., 2005; Dellovade et al., 2001; Henderson et al., 1999). Figure 4.1 illustrates one example of using video microscopy techniques to examine cell movements in and around the developing VMN in vitro. Mice with the SF-1 promoter driving enhanced green fluorescent protein (SF-1/eGFP) (Stallings et al., 2002) and mice with the Thy-1 promoter driving yellow fluorescent protein YFP (Thy-1/YFP) (Feng et al., 2000) expression both provide superior ways to visualize the migration of neurons within the hypothalamus (Tobet et al., 2003). Thy-1 is a neural cell adhesion glycoprotein expressed during development that is evident in cells of the hypothalamus as early as E13. Methods involving transgenic mice can be used to follow the migration path of specific cell populations in vitro and in vivo as well as experimentally manipulate the migratory process, determining cues that are essential to cell movements.

4.3   RADIAL MIGRATION

One way to examine brain compartments and pathways of migration in the hypothalamus is to consider the distribution of radial glial processes through the region at different ages. The processes of radial glia have been hypothesized to function as guides for neuronal migration in the developing brain (Rakic, 1972; 1990; Levitt and Rakic, 1980; Dupouey et al., 1985). Several studies have concentrated on radial glial cells related to the diencephalon (e.g., Levitt and Rakic, 1980; Seress, 1980; Tobet and Fox, 1989).

Neurons and glia are distinct classes of neurons that during early development arise from the epithelium of the neural tube. As the tube expands and the cerebral vesicles enlarge, the undifferentiated cells elongate and the processes remain attached to the pial surface of the neural tube. As these cells elongate they maintain a radial orientation and an early precursor state. These cells can later become neurons or glial cells, however the radial orientation provides a guide for newly "born" cells during the migratory phase of development.

Radial migration in the diencephalon is part of the process in moving cells from the proliferative zone along the third ventricle to occupy nuclei within the thalamus and hypothalamus. As described above for the region of the telencephalon, radial migration occurs along radial cells whose cell bodies are found within the subventricular zone and extend to the pial surface, acting as a scaffold for migrating neurons. Evidence of radial migration is visualized by imaging of neurons, which are likely adhering to the radial fiber. Figure 4.3 illustrates this concept, terminally mitotic neurons are found adjacent to radial fibers in the region of the anterior hypothalamus. By examining the direction of movement in hypothalamic brain slices that included the VMN, the orientation of motion tracked the orientation of glia more than 65% of the time through the center of the developing nucleus (Dellovade et al., 2001). The orientation of migration was less likely to track the orientation of radial glia closer to the proliferative zone and also toward the lateral edge of the developing nuclear group.

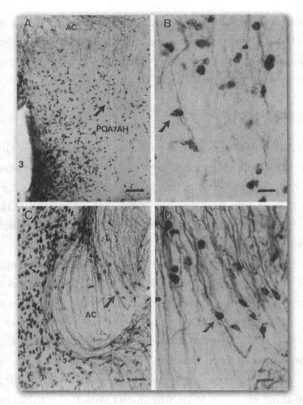

Figure 4.3: BrdU labeling of terminally mitotic cells within the region of the POA/AH. Panels B and D are magnified images of the neurons in close proximity to radial fibers (adapted from Tobet et al., 1995).

The migration of many neural precursor cells is accomplished by somal translocation (Morest et al., 2003). This method of cell movement is accomplished by the movement of the cell body through elongation and contraction of the nucleus. Active rearrangement of the cytoskeleton is an important part of this process. Factors that influence migration can contribute change the directionality, duration, and speed of movement through influencing the assembly and degradation of microtubules. Factors such as reelin promote migration in the cortex through the promotion of microtubule assembly (Meseke et al., 2013). Many neurons in the hypothalamus migrate radially away from ventricular zones guided by radial glial processes and tangential to such fibers, often along neuronal processes. In Figure 4.3, labeled cells are migrating towards the POA/AH in the developing mouse brain. The arrows in the image indicate cells that are opposed to radial fibers. The use of confocal microscopy in more recent studies has demonstrated that this opposition of cell and fiber is likely a physical connection between the cell body and the radial fiber (McClellan et al., 2008).

4.4   SEX DEPENDENCE

One of the first indications that cell migration in the hypothalamus might be sex and hormone-dependent came from the observation of differences in immunoreactivity in radial glial fibers during early development in rats (Tobet and Fox, 1989). The study used a monoclonal antibody that recognized a carbohydrate epitope present on many biochemical entities. It was difficult to clearly identify the molecular basis for the sex difference in immunoreactivity, although a candidate was noted (Tobet et al., 1991). Since that time there have only been a handful of studies to examine directly the influence of sex or gonadal steroids on cell movements in the brain (Tobet et al., 1994; Williams et al., 1999; Henderson et al., 1999, Knoll et al., 2007). Nonetheless, the evidence favors hormonal regulation of cell movements in selective locations within the hypothalamus. In addition, there are a growing number of examples of sex differences in cell positions that provide additional circumstantial evidence for sex differences and steroid hormone influences on cell migration in the hypothalamus. These include the position of cells containing immunoreactive estrogen receptor (ER)-β in the rat AVPV (Orikasa et al., 2002), the position of cells containing immunoreactive neuronal nitric oxide synthase (nNOS) in the preoptic area of estrogen receptor knockout mice (Scordalakes et al., 2002), and data on cells containing immunoreactive ER-β in the mouse as well as GABA receptor subunits and calbindin (Wolfe et al., 2005; Edelmann et al., 2007; Büdefeld et al., 2008). Recently, a sex difference also was noted in the position of cells containing immunoreactive ER-β in and around the paraventricular nucleus of the hypothalamus, particularly as related to the absence of GABAB receptor function (McClellan et al., 2010).

CHAPTER 5

The Long and Winding Road of GnRH Neuronal Migration

Most of the predicted pathways for cell migration in the hypothalamus and other brain regions are aligned with the pattern of radial glial processes (Edwards et al., 1990). However, this does not account for cells migrating in the rostral-caudal dimension, such as those containing the decapeptide gonadotropin-releasing hormone (Schwanzel-Fukuda and Pfaff, 1989; Wray et al., 1989; Tobet et al., 1993a). Neuronal migration that is not associated with the orientation of radial glial processes has been postulated to relate to axonal guidance cues (Rakic, 1990), or non-contact-mediated chemical cues (Book and Morest, 1990), An additional view of glial cells in development is that they provide boundaries, preventing cells or growth cones from entering different compartments (Steindler et al., 1989; Hutchins and Casagrande, 1990; Silver et al., 1993).

Neurons that synthesize gonadotropin-releasing hormone (GnRH) travel a long distance relative to their size! GnRH was first sequenced as a decapeptide in 1971 earning Andrew Schally and Roger Guillemin Nobel prizes in 1977. Over the last 40+ years, different forms of GnRH have been found in multiple species and often more than one form in the same species. The form that controls the pituitary release of gonadotropins in mammals is referred to as mGnRH or GnRH-I. A surprise in 1989 was the finding that the GnRH neurons that regulate pituitary function are born outside of the CNS in the olfactory periphery. GnRH neurons are born in the olfactory placodes (or their vomeronasal organ derivative) and migrate over the nasal septum and across the cribriform plate into the basal forebrain where most turn toward the hypothalamus. Their goal is to send processes to the capillaries of the median eminence that form the first portion of a portal circulation that feeds high concentrations of hypothalamic factors to the pituitary. The cues that drive this unique migration include soluble factors that can work at a distance as well as factors that relate GnRH neurons to guiding fibers. One soluble secreted factor is gamma-aminobutyric acid (GABA), normally thought of as an inhibitory neurotransmitter in adult animals. Maternal treatments that influence GABA alter the migration of GnRH neurons in fetuses, but to determine if the treatments work directly in fetuses in vitro methods were needed (Bless et al., 2000). Another soluble secreted factor is glutamate (Simonian & Herbison, 2001), so substances that are neurotransmitters in the long run may act as chemokinetic factors in development. Kallmann's syndrome is characterized by hypothalamus based infertility and anosmia. One common cause for both symptoms can arise if peripheral olfactory fibers fail to reach the brain or reach the CNS, but abnormally. One signaling system that influences the ability of olfactory fibers that guide GnRH neurons to find their normal paths is the netrin-1 signaling system that can use either the receptors Dcc or Unc5h3 (Schwarting et al., 2004). By looking at carbohydrate markers one can get a feel for the heterogeneity of adhesion mechanisms from which GnRH neurons may take advantage. Even a soluble signaling system, like GABA, may influence the relationship of GnRH neurons to their fibers. Treatments that influence GnRH neurons in development also influence reproductive functions in adults. Another way to show that was using transgenic animals, in which the synthetic enzyme for GABA—glutamic acid decarboxylase—was driven by the GnRH promoter (Heger et al., 2003). Again, GnRH neuron positions were altered. Live viewing suggests that when the fibers are cut distal to the locations of migrating neurons, the neurons still alter their behavior suggesting that there is significant communication between neurons and guiding fibers (see Figure 5.1). The genes referred to as Kal 1 and 2: anosmin and FGF receptor are the most prominently associated with the syndrome now (for review, see Wierman et al., 2011).

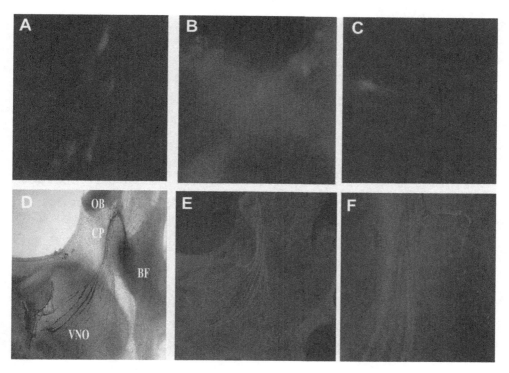

Figure 5.1: Images show that live GFP labeled GnRH neurons maintain a similar morphology in vitro as that seen post hoc using immunohistochemistry. In panels A-C, images show live GFP containing GnRH neurons in 250 μm-thick slices through the nasal compartment (A), cribriform plate (B), and forebrain (C). The image in panel D shows almost the entire GnRH migratory route in a 250 μm-thick slice from an E13 mouse head with immunoreactive GFP. The images in panels E and F (higher magnification) show immunoreactive peripherin, which demonstrates the fiber guiding pathway for GnRH neurons and an intact cribriform plate in the same E13 mouse head slice as that shown in Panel D.

CHAPTER 6

Formation of Cell Groups

Following neuronal divisions along proliferative zones, many cells migrate to their hypothalamic locations and form cell groups (albeit not GnRH neurons). BrdU studies in the mouse have shown that many cells in the developing hypothalamus undergo final mitotic divisions as early as E10, while some cells might not undergo final divisions until E15/E17. Using Nissl stains, many hypothalamic cell groups do not appear as a distinct collection of cells on either side of the third ventricle until around E18 and E19 in rats (Coggeshall, 1964; Hyyppä, 1969), E16 and E17 in mice (Schambra et al., 1991; Tobet et al., 1999), and gestational weeks 9–15 in the human (Koutcherov et al., 2002). The apparent organization of cells into a "nuclear structures" involves both the arrival and arrangement of cells, and the development of surrounding fibers that cause the nuclei to appear more densely cellular than surrounding regions. Interestingly, cells identified by phenotype (e.g., ERα) often can be found in their correct positions before the boundaries of several nuclear groups can be discerned by Nissl stains, e.g., VMN (Tobet et al., 1999) and PVN (McClellan et al., 2010). The successful formation of cell groups involves more than just migration of post-mitotic neurons —other factors present in the developing hypothalamus play a role in this.

6.1   GRADIENTS

The importance of gradients is well established throughout developmental biology. Early embryonic development requires expression patterns of genes to establish anterior/posterior and dorsal/ventral axis formation. Gradients work to cause changes in patterning, allowing for variability in development. A cell's ability to respond to a gradient is variable, with cell populations responding differentially to a threshold of expression. In the hypothalamus, expression patterns of secreted molecules indicate that gradients may be important in establishing the boundaries of nuclear groups (see supplemental video 2 illustrating how the release of a secreted factor can influence nearby cells). The role of neurotransmitters in development is not clearly understood before the development of synaptic connectivity, and these comprise one category of factors that may provide gradients for cells that are competent to respond to the signals. One example of this is seen in the distribution of the neurotransmitter GABA. GABA is found in cells throughout the hypothalamus but its synthetic enzymes glutamic acid decarboxylases (GAD) are excluded from cell bodies within the developing VMN and PVN (Figure 6.1). The patterning of GABA and its synthesizing enzymes can be visualized through immunohistochemistry. The early expression and compartmentalization of GABA implies that it may be important in providing boundary signals for nuclear groups. The gene Rax was found to be critical to the formation of the VMN through induction of multiple cell phenotypes. Interestingly, mice lacking functional Rax exhibit a change in the pattern of GAD immunoreactivity. GAD is found within cells of the VMN in contrast to its normal pattern of expression in cells surrounding the nucleus. According to this study, Rax deficient mice no longer have expression of certain cell phenotypes within the VMN (Lu et al., 2013). It remains to be determined if this is due to a lack of expression or a change in migration and arrangement of cells.

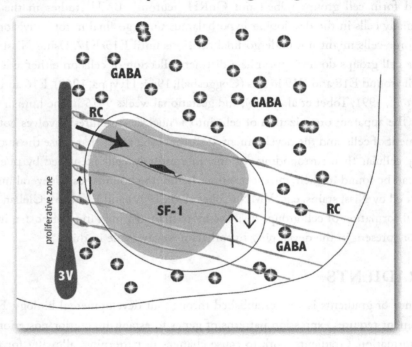

Figure 6.1: Schematic representation showing migration from the proliferative zone along the third ventricle (3V) to fill the VMN. Cells migrate along fibers of radial cells (RC) which have their cell bodies located adjacent to the 3V and long processes that extend to the pial surface of the brain. SF-1 and GABA are both important factors in this migratory process with SF-1 acting intrinsically within the cells of the VMN and GABA acting extrinsically as a signaling molecule to the migrating cells. Migration occurs dorsal/ventral close to the ventricle, in a radial pattern throughout most of the VMN, and dorsal/ventral in the lateral edges of the nucleus (as indicated by the arrows; adapted from McClellan et al., 2006).

GABA exerts a similar role in the cortex through its regulation of migrating cells (Behar et al., 1996). How GABA influences the formation of nuclear groups is a difficult question. The presence of multiple receptor types (GABAA, GABAB, and GABAC) in the hypothalamus indicates that it could be influencing cells in multiple ways. GABA signaling may alter the likelihood of migration (Behar et al., 1996), the speed of migration (McClellan et al., 2008; Figure 6.2) or may influence interactions between migrating cells and radial cell fibers (Bless et al., 2005). Antagonists of GABA receptors caused an increase in the speed of migrating neurons within the VMN but not ARC nuclei (Figure 6.2). Interestingly, a recent study suggests that GABA influences the developing vasculature in the PVN (Frahm et al., 2012) providing another route to altering neuronal development. It is likely that other secreted molecules also provide gradient information for the development of cell groups within the hypothalamus.

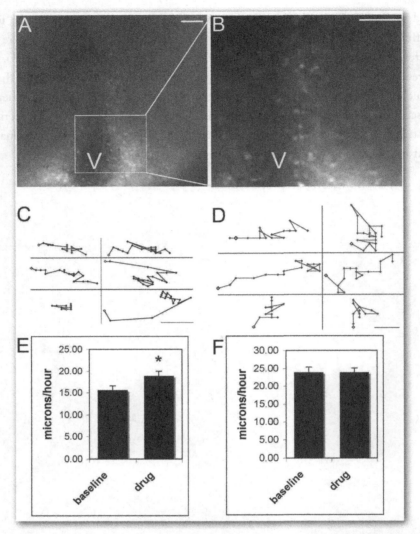

Figure 6.2: Live video microscopy was used for the collection of data examining the effects of GABA antagonists on cell movement within the area of the VMN. The entire region of the hypothalamus that includes the VMN and ARC (A) has fluorescently labeled cells located near the ventricles (V). The boxed region represents the region magnified in (B; scale bars = 100 μm). Migrating neurons did not follow strictly radial or dorsal/ventral movement patterns, rather they changed directions or speeds numerous times within any given time period. Scatter plots for VMN (C) and ARC (D) cells represent sample neurons tracked over 90 min time periods (scale bars = 5 μm) with baseline shown on the left and drug response on the right. In the VMN (C,E) the baseline speeds were significantly slower than those observed following drug treatment. Other nuclei, such as the ARC, were unaffected (B,F). Adapted from McClellan et al., 2008.

6.2   INDIVIDUAL GENES

In at least one case, a single gene factor might be particularly important for the formation of a cell group in the hypothalamus. Steroidogenic Factor-1 (SF-1) is a transcription factor that is essential for the development of the VMN and is restrictively expressed within a subset of cells in the developing VMN. Genetic disruption of SF-1 alters the distribution of surrounding cells that do not express SF-1, i.e., the region is rearranged. Along with the molecular rearrangement of cells, mice lacking SF-1 have physiological changes as well. SF-1 is essential to normal pituitary gonadotrope function, exhibiting a decrease in LH and FSH production in mice lacking SF-1 (Zhao et al., 2001). SF-1 also plays a role in energy balance and when disrupted causes obesity in mice (Majdic et al., 2002). Obesity in SF-1 knockout mice is likely due to the role estrogen plays in SF-1 positive cells (Xu et al., 2011). The loss of ERα in SF-1 positive cells of the VMN results in obesity and decreased metabolism.

A similar case is seen in the development of the PVN. Sim-1 expression is critical to proper development of the PVN and the SON. A partial loss of Sim-1 during hypothalamic development in mice, results in decreased levels of vasopressin and oxytocin neurons by over 50%. The partial loss of Sim-1 also results in a decrease of neuronal projections to the dorsal vagal complex by 70% (Duplan et al., 2009). In certain cases a single gene may regulate the placement or positioning of a certain cell population. Hes1 is a Notch signaling gene required for normal pituitary formation. In Hes1 null mice cells of the PVN and SON, positive for immunoreactive vasopressin, are misplaced from their normal location. This misplacement did not accompany an increase/decrease in cell death or cell proliferation indicating that the altered positioning may be due to a migratory deficit in cells that project to the pituitary gland (Aujla et al., 2011).

The link between forming a functional group, such as the VMN, and proper regulation of behaviors is important to understanding the need for proper migration and arrangement of cells. Changes in the expression of single genes can cause changes in the projections from either the PVN (Sim-1 or Hes1) or VMN (SF-1). The assumption that the organization of the hypothalamus is critical to its function is strengthened by studies such as those described above looking at rearrangement of cells and changes in physiology or behavior.

6.3   NEURONAL SPECIFICATION

There is a tremendous diversity of neurons within the hypothalamus that can be characterized on the basis of chemical phenotype in additional to morphological characteristics. Sometimes these phenotypes do not become obvious until after cells have been through their migratory process (e.g., estrogen receptor identity may develop late relative to other processes; Tobet et al., 1993; Henderson et al., 1999). On the other hand, sometimes cell phenotype develops early and phenotypically identified cells such as those containing GnRH may be followed during the migratory process

(Bless et al., 2005). Each cell group of the hypothalamus contains subsets of cells with identifiable phenotypes. The PVN contains many cell phenotypes accounting for the many functions associated with this region (see Figure 2.2).

Identifying neuronal phenotypes and when they emerge is important to understanding hypothalamic development. Identification of multiple phenotypes within a cell group can help identify changes that are often missed. For example, a Sim-1 deficiency as discussed earlier causes a decrease in levels of AVP and Oxytocin expression but does not change levels of CRH and Thyrotropin Releasing Hormone (TRH; Duplan et al., 2009). A loss of gene expression or axonal connectivity as seen in these mice is not likely to be seen in gross examination of the hypothalamus, histochemical techniques allow examination of individual cell phenotypes and provide a method for investigating location and timing of neuronal specification.

CHAPTER 7

Connectivity

Proper development of the hypothalamus requires interactions of neurons within the hypothalamus and to other areas of the brain. Axon guidance and the establishment of synaptic connections is, for the most part, a final stage of development. It occurs when cells send axonal projections to various other brain targets. These can be intrahypothalamic or extrahypothalamic connections.

The hypothalamus is subdivided in such a way that connectivity can be observed between nuclear groups. The connections between groups illustrate the complexity of how behaviors are regulated. Intrahypothalamic connections exist between groups regulating similar behaviors. For example, the VMN, ARC, and PVN are key players in the regulation of neuroendocrine function, specifically in metabolism and energy homeostasis. These three nuclei are interconnected, however, those connections become even more complex as specific cell identities are examined. Axons from the VMN influence neurons of the PVN by projecting to areas ventral and dorsal to the PVN (Ter Horst et al., 1986). In contrast, input from parvocellular neurons of the PVN synapse with neurons in the medial region of the VMN. Intrahypothalamic connections are prevalent in capsules surrounding nuclear groups. Labeling of fiber tracts surrounding the VMN illustrates an extensive dendritic capsule that provides a point of connectivity for afferent neurons (Flanagan-Cato et al., 2011). These dendritic capsules are receiving input from other nuclei within the hypothalamus and outside of the region. In addition to complex connections between nuclei, some hypothalamic groups also contain abundant intranuclear connections. The VMN is one nuclear group with abundant intranuclear projections as cells from the ventrolateral region connect with more dorsomedial cells (TerHorst et al., 1986). The cells within the VMN project to the dendritic capsule surrounding the cell-rich area of the nucleus. Altered rearrangement of projections from the PVN is impacted by the rearrangement of cells within the VMN (Büdefeld et al., 2011). Figure 7.1 is a simplified schematic of the intrahypothalamic and extrahypothalamic connections of the VMN.

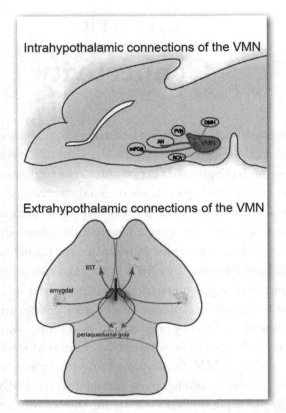

Figure 7.1: Connections between hypothalamic nuclei and connections with other brain regions are critical for finishing the job of building a functional hypothalamus. The development of connections is complex as nuclear groups within the hypothalamus establish connections between groups, receive afferent inputs from other areas, and send efferent output throughout the brain. As an example, the VMN sends and receives input from the dorsomedial hypothalamus (DMH), PVN, AH, and medial POA. Extrahypothalamic inputs extend to the BNST, the amygdala, regions of the brainstem (periaqueductal gray), and others. In addition to the extensive number of direct input, indirect connections are also found throughout the brain. This schematic greatly simplifies the extensive amount of information that is sent to/from the hypothalamus and the importance of establishing connections early on.

The nuclei of the hypothalamus have an extensive network of extrahypothalamic connections that allow the region to exert control on numerous physiological systems. The extrahypothalamic projections can be further subdivided into afferent and efferent fibers. Afferent fibers come from neighboring areas such as the preoptic area, thalamus, epithalamus, and amygdala while other axonal connections are traveling longer distances from the bed nucleus of the stria terminalis (BNST), areas of the midbrain and medulla, and the cerebral cortex (Canteras et al., 1994; Fahrbach et al., 1989). Efferent projections leaving the hypothalamus connect to locations nearby and distant as well. The amygdala, preoptic area, thalamus, brainstem nuclei, and cortical structures are among a few (Saper et al., 1976). A recent study by Hahn and Swanson emphasized how extensive the connections of a single hypothalamic area can be. The lateral hypothalamic area connects with nearly every major division of the cerebrum and brainstem (Hahn and Swanson, 2010).

Information on when projections are established is not well known. Early studies indicate that connections are established within the first postnatal week of the rodent brain (Choi et al., 2005). Studies aimed at analyzing factors involved in directing the formation of connections examine rodents during this first postnatal week of life. The transcription factor Lhx6 is involved in the establishment of connections between the amygdala and the VMN and that these connections may be involved in the regulation of reproductive behaviors (Choi et al., 2005). The development of circuits involved in feeding has been one area of focus in establishing a timeline for circuit development. Efferent projections from the ARC involved in appetite regulation are established during the second postnatal week in the mouse (Bouret et al., 2004). These efferent fibers project to the dorsomedial hypothalamus, lateral hypothalamic area, and PVN. This same study also revealed that a deficiency in leptin signaling resulted in a delay in the formation of ARC projections. It is proposed that leptin acts as a neurotrophic factor in promoting the connections of neurons involved in energy balance (Bouret and Simerly, 2007). More recently the targets of leptin-dependent fibers from the ARC to the PVN were identified as preautonomic (Bouyer and Simerly, 2013).

The development of projections to (Hutton et al., 1998) and from (Polston & Simerly, 2006) the anteroventral periventricular preoptic area (AVPV) has also been examined. Interestingly, the projection of the bed nucleus of the stria terminalis to AVPV is sex and hormone dependent being driven by testosterone exposure in males. That projection does not develop until the second postnatal week. Projections form the AVPV develop early when they are rostrally oriented, particularly toward GnRH neurons where the connections are made prior to birth. For the more numerous caudal projections, they develop over the second and third postnatal weeks.

CHAPTER 8

Rebuilding the Hypothalamus

Neurogenesis does not simply occur during prenatal life or during times of learning. A growing field of research is aimed at looking at the evidence and significance of adult neurogenesis. There has been consensus regarding neurogenesis in adulthood in the hippocampus and olfactory bulb across a number of species. Recent years have seen an upsurge in reports concerning neurogenesis in additional regions. Some, like the cortex, have been controversial. Others, like the hypothalamus, are relatively recent to the party (Migaud et al., 2010; Lee et al., 2012). A major question remains as to the function of such new neurons in the hypothalamus (Lee & Blackshaw, 2012). The generation of new neurons by itself does not have much functional significance. However, the incorporation of newly generated neurons into the elaborate circuitry of the hypothalamus would indicate a role in regulating physiology or behavior. The function frequently considered recently is related to feeding and energy balance. Newly generated neurons of the arcuate nucleus express orexigenic and anorexigenic peptides similar to established cell phenotypes (Kokoeva et al., 2005; Lee et al., 2012). A role in puberty has also been suggested (Ahmed et al., 2008).

A growing body of data is suggesting that there are neurogenic niche(s) in the hypothalamus in adulthood. The strongest candidates for providing a proliferating population are currently tanycytes, but it is not clear if that can explain all of the incidences of new neurons in adults (Lee et al., 2012). The mechanism for adult neurogenesis in the hypothalamus is not well understood. Growth factors may be involved; an increase in insulin-like growth factor 1 (IGF-1) was shown to stimulate neurogenesis in the rat hypothalamus. The area of proliferation was seen along the subventricular zone and in the layer of tanycytes (Perez-Martin M, 2010).

The regulation of adult neurogenesis in the hippocampus can be modified through behavior and lifestyle. Enriched learning environments, exercise, and hormones are some of the factors shown to influence the rate of adult neurogenesis (Migaud et al., 2010). Effects of adult neurogenesis on the hippocampus are intriguing. The extensive network of connections between the hippocampus and hypothalamus may lead to a functional role for neurogenesis within the stress axis.

CHAPTER 9

Conclusion

This discussion started over the thought of building a hypothalamus. That thought took us through a number of fundamental processes that stretched from generating the cellular building blocks (neurogenesis) to determining where they go (migration), whether they should be kept (cell death), what they should become (cell fate), and finally who they should talk to (connectivity). These processes rely on the same fundamental mechanisms as other brain regions including cytoskeletal dynamics, cellular adhesion, and cell-to-cell signaling. The hypothalamus, however, adds special flavor in two ways. First, the hypothalamus is particularly attuned to hormonal signaling. Numerous cells attend to hormonal signaling from the periphery (e.g., the presence of steroid hormone receptors in many cell nuclei) and numerous cells have direct access to secrete important peptides into the peripheral circulation (e.g., GnRH or vasopressin). Second, the hypothalamus is particularly replete with sex differences in neural structure and function. Unraveling the development of these structures and functions provide important clues for potential enhancements, rebuilding, or repair.

Bibliography

Acampora, D., Postiglione, M.P., Avantaggiato, V., Di Bonito, M., Simeone, A., 2000. The role of Otx and Otp genes in brain development. *Int. J. Dev. Biol.* 44, pp. 669–677. 8

Ahmed, EI., Zehr, J.L., Schulz, K.M., Lorenz, B.H., DonCarlos, L.L., Sisk, C.L., 2008. Pubertal hormones modulate the addition of new cells to sexually dimorphic brain regions. *Nat. Neurosci.* 11, pp. 995-7. 39

Altman, J., Bayer, S.A., 1978. Development of the diencephalon in the rat. I. Ontogeny of the specialized ventricular linings of the hypothalamic third ventricle. *J. Comp. Neurol.* 182, pp. 945-71. DOI: 10.1002/cne.901820511

Altman, J., Bayer, S.A., 1978. Development of the diencephalon in the rat. II. Ontogeny of the specialized ventricular linings of the hypothalamic third ventricle. *J. Comp. Neurol.* 182, pp. 973-93. DOI: 10.1002/cne.901820512. 6

Altman, J., Bayer, S.A., 1978. Development of the diencephalon in the rat. III. Ontogeny of the specialized ventricular linings of the hypothalamic third ventricle. *J. Comp. Neurol.* 182, pp. 995-1015. Neuron Development `http://neurondevelopment.org/`. DOI: 10.1002/cne.901820513 6

Altman, J., Bayer, S.A., 1986. The development of the rat hypothalamus. *Adv. Anat. Embryol. Cell Biol.* 100, pp. 1-178. 7, 9

Alvarez-Bolado G., Rosenfeld, M.G., Swanson, L.W., 1995. Model of forebrain regionalization based on spatiotemporal patterns of POU-III homeobox gene expression, birthdates, and morphological features. *J. Comp. Neurol.* 355, pp. 237–295. DOI: 10.1002/cne.903550207

Anthony, T.E., Klein, C., Fishell, G., Heintz, N., 2004. Radial glia serve as neuronal progenitors in all regions of the central nervous system. *Neuron.* 41, pp. 881-90. 11

Armstrong, W.E., Warach, S., Hatton, G.I., McNeill, T.H., 1980. Subnuclei in the rat hypothalamic paraventricular nucleus: a cytoarchitectural, horseradish peroxidase and immunocytochemical analysis. *Neuroscience*, 5, pp. 1931-58. 5

Aujla, P.K., Bora, A., Monahan, P., Sweedler, J.V., Raetzman, L.T., 2011. The Notch effector gene Hes1 regulates migration of hypothalamic neurons, neuropeptide content and axon targeting to the pituitary. *Dev. Biol.* 353, pp. 61-71. DOI: 10.1016/j.ydbio.2011.02.018. 18, 33

Bayer, S.A., 1979. The development of the septal region in the rat. I. Neurogenesis examined with 3H-thymidine autoradiography. *J. Comp. Neurol.* 183, pp. 89-106. DOI: 10.1002/cne.901830108 6

Bayer, S.A., Altman, J., 1987. Development of the preoptic area: time and site of origin, migratory routes, and settling patterns of its neurons. *J. Comp. Neurol.* 265, pp. DOI: 10.1002/cne.902650106 65-95. 6

Behar, T.N., Li, Y.X., Tran, H.T., Ma, W., Dunlap, V., Scott, C., Barker, J.L., 1996. GABA stimulates chemotaxis and chemokinesis of embryonic cortical neurons via calcium-dependent mechanisms. *J. Neurosci.* 16, pp. 1808-18. 30, 31

Behar, T.N., Dugich-Djordjevic, M.M., Li, Y.X., Ma, W., Somogyi, R., Wen, X., Brown, E., Scott, C., McKay, R.D., Barker, J.L., 1997. Neurotrophins stimulate chemotaxis of embryonic cortical neurons. *Eur. J. Neurosci.* 9 pp. 2561-70. 10

Berghard A., Hägglund A.C., Bohm S., Carlsson L., 2012. Lhx2-dependent specification of olfactory sensory neurons is required for successful integration of olfactory, vomeronasal, and GnRH neurons. *FASEB J.* 2012 8, pp. DOI: 10.1096/fj.12-206193. 3464-72. 8

Blaschke, A.J., Staley, K., Chun, J., 1996. Widespread programmed cell death in proliferative and postmitotic regions of the fetal cerebral cortex. *Development.* 122, pp. 1165-74. 13

Bless, E.P., Westaway, W.A., Schwarting, G.A., Tobet, S.A., 2000. Effects of gamma-aminobutyric acid(A) receptor manipulation on migrating gonadotropin-releasing hormone neurons through the entire migratory route in vivo and in vitro. *Endocrinology*, 141, pp. 1254-62. DOI:10.1210/en.141.3.1254. 25

Bless, E.P., Walker, H.J., Yu, K.W., Knoll, J.G., Moenter, S.M., Schwarting, G.A., Tobet, S.A., 2005. Live view of gonadotropin-releasing hormone containing neuron migration. *Endocrinology.* 146, pp. 463-8. DOI:10.1210/en.2004-0838. 20, 31, 34

Book, K.J., Morest, D.K., 1990. Migration of neuroblasts by perikaryal translocation: role of cellular elongation and axonal outgrowth in the acoustic nuclei of the chick embryo medulla. *J. Comp. Neurol.* 297, pp. 55-76. 25

Bouret, S.G., Simerly, R.B., 2004. Minireview: Leptin and development of hypothalamic feeding circuits. *Endocrinology* 145, pp. 2621-6. DOI: 10.1210/en.2004-0231. 5

Bouret, S.G., Draper, S.J., Simerly, R.B., 2004. Formation of projection pathways from the arcuate nucleus of the hypothalamus to hypothalamic regions implicated in the neural control of feeding behavior in mice. *J. Neurosci.* 24, pp. 2797-805. DOI: 10.1523/JNEUROSCI.5369-03.2004. 37

Bouret, S.G., Simerly, R.B., 2007. Development of leptin-sensitive circuits. *J. Neuroendocrinol.* 19, pp. . 575-82. 37

Bouyer, K., Simerly, R.B., 2013. Neonatal leptin exposure specifies innervation of presympathetic hypothalamic neurons and improves the metabolic status of leptin-deficient mice. *J. Neurosci.* 33, pp. 840-51. DOI: 10.1523/JNEUROSCI.3215-12.2013 37

Büdefeld, T., Grgurevic, N., Tobet, S.A., Majdic, G., 2008. Sex differences in brain developing in the presence or absence of gonads. *Dev. Neurobiol.* 68, pp. 981-95. DOI: 10.1002/dneu.20638. 23

Büdefeld, T., Tobet, S.A., Majdic, G. 2011. Altered position of cell bodies and fibers in the ventro-medial region in SF-1 knockout mice. *Exp. Neurol.* 232, pp. 176-84. DOI: 10.1016/j.expneu-rol.2011.08.021. 18, 35

Buss, R.R., Sun, W., Oppenheim, R.W., 2006. Adaptive roles of programmed cell death during nervous system development. *Annu. Rev. Neurosci.* 29, pp. 1-35. DOI: 10.1146/annurev.neuro.29.051605.112800. 13

Canteras N.S., Simerly R.B., Swanson L.W., 1994. Organization of projections from the ventro-medial nucleus of the hypothalamus: a Phaseolus vulgaris-leucoagglutinin study in the rat. *J. Comp. Neurol.* 348, pp. 41-79. DOI: 10.1002/cne.903480103. 5,

Caqueret, A., Yang, C., Duplan, S., Boucher, F., Michaud, J.L., 2005. Looking for trouble: a search for developmental defects of the hypothalamus. *Horm. Res.* 64, pp. 222-30. 1

Caqueret, A., Boucher, F., Michaud, J.L., 2006. Laminar organization of the early developing ante-rior hypothalamus. *Dev. Biol.* 298, pp. 95-106. DOI: 10.1016/j.ydbio.2006.06.019

Carbone, D.L., Handa, R.J., 2012. Sex and stress hormone influences on the expression and activity of brain-derived neurotrophic factor. *Neuroscience* 12. DOI: 10.1016/j.neurosci-ence.2012.10.073. 10

Caviness, V.S. Jr, Goto, T., Tarui, T., Takahashi, T., Bhide, P.G., Nowakowski, R.S. 2003. Cell out-put, cell cycle duration and neuronal specification: a model of integrated mechanisms of the neocortical proliferative process. *Cereb. Cortex.*, 13, pp. 592-8. 20

Choi, G.B., Dong, H.W., Murphy, A.J., Valenzuela, D.M., Yancopoulos, G.D., Swanson, L.W., Anderson, D.J., 2005. Lhx6 delineates a pathway mediating innate reproductive behaviors from the amygdala to the hypothalamus. *Neuron.* 46, 647-60. 37

Chung, W.C., Swaab, D.F., De Vries, G.J., 2000. Apoptosis during sexual differentiation of the bed nucleus of the stria terminalis in the rat brain. *J. Neurobiol.* 43, pp. 234-43. DOI: 10.1002/(SICI)1097-4695(20000605)43:3<234::AID-NEU2>3.0.CO;2-3. 13

Chung, W.C., Tsai, P.S., 2010. Role of fibroblast growth factor signaling in gonadotropin-releasing hormone neuronal system development. *Front. Horm. Res.* 39, pp. 37-50. DOI: 10.1159/000312692. 8

Creps, E.S., 1974. Time of neuron origin in preoptic and septal areas of the mouse: an autoradiographic study. *J. Comp. Neurol.* 157, pp. 161-243. DOI: 10.1002/cne.901570205. 6

Coggeshall, R.E., 1964. A study of diencephalic development in the albino rat. *J. Comp. Neurol.* 122, pp. 241-69. 29

Czupryn, A., Zhou, Y.D., Chen, X., McNay, D., Anderson, M.P., Flier, J.S., Macklis, J.D., 2011. Transplanted hypothalamic neurons restore leptin signaling and ameliorate obesity in db/db mice. *Science.* 334, pp. 1133-7. DOI: 10.1126/science.1209870

Davis, A.M., Seney, M.L., Stallings, N.R., Zhao, L., Parker, K.L., Tobet, S.A., 2004. Loss of steroidogenic factor 1 alters cellular topography in the mouse ventromedial nucleus of the hypothalamus. *J. Neurobiol.* 60, pp. 424-36. DOI: 10.1002/neu.20030. 13

de Vries, G.J., Jardon, M., Reza, M., Rosen, G.J., Immerman, E., Forger, N.G., 2008. Sexual differentiation of vasopressin innervation of the brain: cell death versus phenotypic differentiation. *Endocrinology* 149, pp. 4632-7. DOI: 10.1210/en.2008-0448

Dellovade, T.L., Young, M., Ross, E.P., Henderson, R., Caron, K., Parker, K., Tobet, S.A., 2000. Disruption of the gene encoding SF-1 alters the distribution of hypothalamic neuronal phenotypes. *J. Comp. Neurol.* 423, pp. 579-89. 18

Dellovade, T.L., Davis, A.M., Ferguson, C., Sieghart, W., Homanics, G.E., Tobet, S.A. 2001. GABA influences the development of the ventromedial nucleus of the hypothalamus. *J. Neurobiol.* 49, pp. 264-76. DOI: 10.1002/neu.10011. 20, 21

Dupouey, P., Benjelloun, S., Gommes, D., 1985. Immunohistochemical demonstration of an organized cytoarchitecture of the radial glia in the CNS of the embryonic mouse. *Dev Neurosci.* 7, pp. 81-93. Doi: 10.1159/000112279. 21

Duplan, S.M., Boucher, F., Alexandrov, L., Michaud, J.L., 2009. Impact of Sim1 gene dosage on the development of the paraventricular and supraoptic nuclei of the hypothalamus. *Eur. J. Neurosci.* 30, pp. 2239-49. DOI: 10.1111/j.1460-9568.2009.07028.x. 33, 34

Edelmann M., Wolfe C., Scordalakes E.M., Rissman E.F., Tobet S. (2007) Neuronal nitric oxide synthase and calbindin delineate sex differences in the developing hypothalamus and preoptic area. *Dev. Neurobiol.* 67, pp. 1371-81. DOI: 10.1002/dneu.20507. 23

Edwards, M.A., Yamamoto, M., Caviness, V.S. Jr., 1990. Organization of radial glia and related cells in the developing murine CNS. An analysis based upon a new monoclonal antibody marker. *Neuroscience*. 36, pp. 121-44. 25

Ever, L., Gaiano, N., 2005. Radial 'glial' progenitors: neurogenesis and signaling. *Curr. Opin. Neurobiol.* 15, pp. 29-33. 10

Fahrbach, S.E., Morrell, J.I., Pfaff, D.W., 1989. Studies of ventromedial hypothalamic afferents in the rat using three methods of HRP application. *Exp. Brain Res.* 77, pp. 221-33. 36

Feng, G., Mellor, R.H., Bernstein, M., Keller-Peck, C., Nguyen, Q.T., Wallace, M., Nerbonne, J.M., Lichtman, J.W., Sanes, J.R., 2000. Imaging neuronal subsets in transgenic mice expressing multiple spectral variants of GFP. *Neuron*. 28, pp. 41-51. 20

Figdor, M.C., Stern, C.D., 1993. Segmental organization of embryonic diencephalon. *Nature*. 363, pp. 630-4. DOI:10.1038/363630a0. 4

Fishell, G., Mason, C.A., Hatten, M.E., 1993. Dispersion of neural progenitors within the germinal zones of the forebrain. *Nature*. 362, pp. 636-8. DOI:10.1038/362636a0. 4

Flament-Durand, J., Brion, J.P., 1985. Tanycytes: morphology and functions: a review. *Int. Rev. Cytol.* 96, pp. 121-55. 11

Flanagan-Cato, L.M., 2011. Sex differences in the neural circuit that mediates female sexual receptivity. *Front. Neuroendocrinol.* 32, pp. DOI: 10.1016/j.yfrne.2011.02.008. 124-36. 35

Forger, N.G., 2009. Control of cell number in the sexually dimorphic brain and spinal cord. *J. Neuroendocrinol.* 21, pp. 393-9. DOI: 10.1111/j.1365-2826.2009.01825.x. 9, 13

Frahm, K.A., Schow, M.J., Tobet, S.A., 2012. The vasculature within the paraventricular nucleus of the hypothalamus in mice varies as a function of development, subnuclear location, and GABA signaling. *Horm. Metab. Res.* 44, pp. 619-24. DOI: 10.1055/s-0032-1304624. 31

Gaiano N., Nye J.S., Fishell G., 2000. Radial glial identity is promoted by Notch1 signaling in the murine forebrain. *Neuron* 26, pp. 395-404. DOI: 10.1016/S0896-6273(00)81172-1. 11

Gelman D., Griveau A., Dehorter N., Teissier A., Varela C., Pla R., Pierani A., Marín O., 2011. A wide diversity of cortical GABAergic interneurons derives from the embryonic preoptic area. *J Neurosci.* 31, pp.16570-80. DOI: 10.1523/JNEUROSCI.4068-11.2011. 4

Gilmore, R.F., Varnum, M.M., Forger, N.G., 2012. Effects of blocking developmental cell death on sexually dimorphic calbindin cell groups in the preoptic area and bed nucleus of the stria terminalis. *Biol. Sex. Differ.* 3, pp. 5-15. DOI: 10.1186/2042-6410-3-5

Hahn, J.D., Swanson, L.W., 2010. Distinct patterns of neuronal inputs and outputs of the jux-taparaventricular and suprafornical regions of the lateral hypothalamic area in the male rat. *Brain Res. Rev.* 64, pp. 14-103. DOI: 10.1016/j.brainresrev.2010.02.002. 37

Hartfuss, E., Galli, R., Heins, N., Gotz, M., 2001. Characterization of CNS precursor subtypes and radial glia. *Dev. Biol.* 229, 15-30. DOI: 10.1006/dbio.2000.9962. 11

Heger, S., Seney, M., Bless, E., Schwarting, G.A., Bilger, M., Mungenast, A., Ojeda, S.R., Tobet, S.A., 2003. Overexpression of glutamic acid decarboxylase-67 (GAD-67) in gonadotropin-re-leasing hormone neurons disrupts migratory fate and female reproductive function in mice. *Endocrinology.* 144, pp. 2566-79. DOI: 10.1210/en.2002-221107. 24

Henderson, R.G., Brown, A.E., Tobet, S.A., 1999. Sex differences in cell migration in the preoptic area/anterior hypothalamus of mice. *J. Neurobiol.* 41, pp. 252-66. DOI: 10.1002/(SICI)1097-4695(19991105)41:2<252::AID-NEU8>3.0.CO;2-W. 20, 23, 33

Heng, J.I., Chariot, A., Nguyen, L., 2010. Molecular layers underlying cytoskeletal remodelling during cortical development. *Trends Neurosci.* 33, pp. 38-47. DOI: 10.1016/j.tins.2009.09.003. 15

Hutchins, J.B., Casagrande, V.A., 1990. Development of the lateral geniculate nucleus: interactions between retinal afferent, cytoarchitectonic, and glial cell process lamination in ferrets and tree shrews. *J. Comp. Neurol.* 298, pp. 113-28. DOI: 10.1002/cne.902980109. 25

Hutton, L.A., Gu, G., Simerly, R.B., 1998. Development of a sexually dimorphic projection from the bed nuclei of the stria terminalis to the anteroventral periventricular nucleus in the rat. *J. Neurosci.* 18, pp. 3003-13. 9, 37

Hyyppä, M., 1969. Differentiation of the hypothalamic nuclei during ontogenetic development in the rat. *Z Anat Entwicklungsgesch.* 129, pp. 41-52. 29

Jacobson, C.D., Gorski, R.A., 1981. Neurogenesis of the sexually dimorphic nucleus of the preoptic area in the rat. *J. Comp. Neurol.* 196, pp. 519-29. DOI: 10.1002/cne.901960313. 11

Jacobson, C.D., Davis, F.C., Gorski, R.A., 1985. Formation of the sexually dimorphic nucleus of the preoptic area: neuronal growth, migration and changes in cell number. *Brain Res.* 353, pp. 7-18.

Jaglin, X.H., Hjerling-Leffler J., Fishell G., Batista-Brito R. 2012 The origin of neocortical nitric oxide synthase-expressing inhibitory neurons. *Front Neural Circuits.* 6, 44. DOI: 10.3389/fncir.2012.00044. 5

Kim K.H., Patel L., Tobet S.A., King J.C., 1999. Gonadotropin-releasing hormone immunoreactivity in the adult and fetal human olfactory system. *Brain Res.* 826, pp. DOI: 10.1016/S0006-8993(99)01271-8 220-9. 10

King, J.C., Rubin, B.S., 1994. Dynamic changes in LHRH neurovascular terminals with various endocrine conditions in adults. *Horm. Behav.* 28, pp. 349-56. DOI: 10.1006/hbeh.1994.1031. 11

Knoll, J.G., Wolfe, C.A., Tobet, S.A., 2007. Estrogen modulates neuronal movements within the developing preoptic area-anterior hypothalamus. *Eur. J. Neurosci.* 26, pp. 1091-9. DOI: 10.1111/j.1460-9568.2007.05751.x. 12, 23

Kokoeva, M.V., Yin, H., Flier, J.S., 2005. Neurogenesis in the hypothalamus of adult mice: potential role in energy balance. *Science* 310, pp. 679-83. DOI: 10.1126/science.1115360. 39

Koutcherov, Y., Mai, J.K., Ashwell, K.W., Paxinos, G., 2002. Organization of human hypothalamus in fetal development. *J. Comp. Neurol.* 446, pp. 301-24. DOI: 10.1002/cne.10175. 1, 29

Koutcherov, Y., Mai, J.K., Paxinos, G., 2003. Hypothalamus of the human fetus. *J. Chem. Neuroanat.* 26, pp. 253-70. DOI: 10.1016/j.jchemneu.2003.07.002. 1

Lechan, R.M., Toni, R., 2008. Functional anatomy of the hypothalamus and pituitary. Chapter 3b in *Neuroendocrinology, Hypothalamus, and Pituitary* (ed., Grossman, A.). Endotext.com. http://www.endotext.org/neuroendo/neuroendo3b/neuroendoframe3b.htm. 1

Lee, D.A., Bedont, J.L., Pak, T., Wang, H., Song, J., Miranda-Angulo, A., Takiar, V., Charubhumi, V., Balordi, F., Takebayashi, H., Aja, S., Ford, E., Fishell, G., Blackshaw, S., 2012. Tanycytes of the hypothalamic median eminence form a diet-responsive neurogenic niche. *Nat. Neurosci.* 15, pp. 700-2. DOI: 10.1038/nn.3079. 39

Lee, D.A., Blackshaw, S., 2012. Functional implications of hypothalamic neurogenesis in the adult mammalian brain. *Int. J. Dev. Neurosci.* Epub ahead of print, Jul 31. 11, 39

Levitt, P., Rakic, P., 1980. Immunoperoxidase localization of glial fibrillary acidic protein in radial glial cells and astrocytes of the developing rhesus monkey brain. *J. Comp. Neurol.* 193, pp. 815-40. DOI: DOI: 10.1002/cne.901930316. 11, 21

Levitt, P., Cooper, M.L., Rakic, P., 1981. Coexistence of neuronal and glial precursor cells in the cerebral ventricular zone of the fetal monkey: an ultrastructural immunoperoxidase analysis. *J. Neurosci.* 1, pp. 27-39.

Lu, F., Kar, D., Gruenig, N., Zhang, Z.W., Cousins, N., Rodgers, H.M., Swindell, E.C., Jamrich, M., Schuurmans, C., Mathers, P.H., Kurrasch, D.M., 2013. Rax is a selector gene for mediobasal hypothalamic cell types. *J. Neurosci.* 33, pp. 259-72. DOI:10.1523/JNEUROSCI.0913-12.2013. 30

Majdic, G., Young, M., Gomez-Sanchez, E., Anderson, P., Szczepaniak, L.S., Dobbins, R.L., McGarry, J.D., Parker, K.L., 2002. Knockout mice lacking steroidogenic factor 1 are a novel genetic model of hypothalamic obesity. *Endocrinology*. 143, pp. 607-14. DOI: 10.1210/en.143.2.607. 18, 33

Majdic, G., Tobet S., 2011. Cooperation of sex chromosomal genes and endocrine influences for hypothalamic sexual differentiation. *Front. Neuroendocrinol.* 32, pp. 137-45. DOI: 10.1016/j.yfrne.2011.02.009. 12

Marín, O., Valiente, M., Ge, X., Tsai, L.H., 2010. Guiding neuronal cell migrations. *Cold Spring Harb. Protoc.* 2, pp. a001834. DOI: 10.1101/cshperspect.a001834.

Markakis, E.A., Swanson, L.W., 1997. Spatiotemporal patterns of secretomotor neuron generation in the parvicellular neuroendocrine system. *Brain Res. Brain Res. Rev.* 24, pp. 255-291. DOI: 10.1016/S0165-0173(97)00006-4. 7

Markakis, E.A., 2002. Development of the neuroendocrine hypothalamus. *Front. Neuroendocrinol.* 23, pp. 257-91. DOI: 10.1016/S0091-3022(02)00003-1. 9

McCabe, M.J., Gaston-Massuet, C., Tziaferi, V., Gregory, L.C., Alatzoglou, K.S., Signore, M., Puelles, E., Gerrelli, D., Farooqi, I.S., Raza, J., Walker, J., Kavanaugh, S.I., Tsai, P., Pittelous, N., Martinez-Barbera, J., Dattani, M.T., 2011. Novel FGF8 mutations associated with recessive holoprosencephaly, craniofacial defects, and hypothalamo-pituitary dysfunction. *J. Clin. Endocrino.l Metab.* 96, pp. 1709-18. DOI: 10.1210/jc.2011-0454. 8

McClellan, K.M., Parker, K.L., Tobet, S., 2006. Development of the ventromedial nucleus of the hypothalamus. *Front. Neuroendocrinol.* 27, pp. 193-209. DOI: 10.1016/j.yfrne.2006.02.002. 14, 17, 30

McClellan, K.M., Calver, A.R., Tobet, S.A., 2008. GABAB receptors role in cell migration and positioning within the ventromedial nucleus of the hypothalamus. *Neuroscience*. 151, pp. 1119-31. DOI: 10.1016/j.neuroscience.2007.11.048. 10, 22, 31, 32

McClellan, K.M., Stratton, M.S., Tobet, S.A., 2010. Roles for gamma-aminobutyric acid in the development of the paraventricular nucleus of the hypothalamus. *J. Comp. Neurol.* 518, pp. 2710-28. DOI: 10.1002/cne.22360. 6, 10, 23, 29

Meseke, M., Cavus, E., Förster, E., 2013. Reelin promotes microtubule dynamics in processes of developing neurons. *Histochem Cell Biol.* 139, pp. 283-97. DOI: 10.1007/s00418-012-1025-1. 22

Michaud, J.L., 2001. The developmental program of the hypothalamus and its disorders. *Clin. Genet.* 60, pp. 255-63. DOI: 10.1034/j.1399-0004.2001.600402.x. 1

Michaud, J.L., Rosenquist, T., May, N.R., Fan, C.M. 1998. Development of neuroendocrine lineages requires the bHLH-PAS transcription factor SIM1. *Genes Dev.* 12, pp. 3264-75. 5

Migaud, M., Batailler, M., Segura, S., Duittoz, A., Franceschini, I., Pillon, D., 2010. Emerging new sites for adult neurogenesis in the mammalian brain: a comparative study between the hypothalamus and the classical neurogenic zones. *Eur. J. Neurosci.* 32, pp. 2042-52. DOI: 10.1111/j.1460-9568.2010.07521.x. 39

Miller M.W., Nowakowski R.S., 1988. Use of bromodeoxyuridine-immunohistochemistry to examine the proliferation, migration, and time of origin of cells in the central nervous system. *Brain Res.* 1, pp 44-52. 7

Misson, J.P., Edwards, M.A., Yamamoto, M., Caviness V.S. Jr., 1988. Identification of radial glial cells within the developing murine central nervous system: studies based upon a new immuno-histochemical marker. *Brain Res. Dev. Brain Res.* 44, pp. 95-108. 11

Moreno-Estellés, M., Díaz-Moreno, M., González-Gómez, P., Andreu, Z., Mira, H., 2012. Single and dual birthdating procedures for assessing the response of adult neural stem cells to the infusion of a soluble factor using halogenated thymidine analogs. *Curr. Protoc. Stem Cell Biol.* Chapter 2: Unit 2D.10. DOI: 10.1002/9780470151808.sc02d10s21. 20

Morest, D.K., Silver, J., 2003. Precursors of neurons, neuroglia, and ependymal cells in the CNS: what are they? Where are they from? How do they get where they are going? *Glia.* 43, pp. 6-18. DOI: 10.1002/glia.10238. 22

Museke, M., Cavus, E., Förster, E., 2013. Reelin promotes microtubule dynamics in processes of developing neurons. *Histochem. Cell Biol.* 139, pp. 283-97. DOI: 10.1007/s00418-012-1025-1

Nauta, Walle J.H., Feirtag, Michael, 1979. The Organization of the Brain. Scientific American, 241, pp. 88-111. DOI: 10.1038/scientificamerican0979-88

Noctor, S.C., Flint, A.C., Weissman, T.A., Dammerman, R.S., Kriegstein A.R., 2001. Neurons derived from radial glial cells establish radial units in neocortex. *Nature*, 6821, pp. 714-20. 11

Orikasa, C., Kondo, Y., Hayashi, S., McEwen, B.S., Sakuma, Y., 2002. Sexually dimorphic expression of estrogen receptor beta in the anteroventral periventricular nucleus of the rat preoptic area: implication in luteinizing hormone surge. *Proc. Natl. Acad. Sci. USA* 99, pp. 3306-3311. DOI: 10.1073/pnas.052707299. 23

Orikasa, C., Sakuma, Y., 2010. Estrogen configures sexual dimorphism in the preoptic area of C57BL/6J and ddN strains of mice. *J. Comp. Neurol.* 518, pp. 3618-29. DOI: 10.1002/cne.22419. 12

Park, D., Xiang, A.P., Zhang L., Mao F.F., Walton, N.M., Choi, S.S., Lahn, B.T., 2009. The radial glia antibody RC2 recognizes a protein encoded by Nestin. *Biochem. Biophys. Res. Commun.* 382, pp. 588-92. DOI: 10.1016/j.bbrc.2009.03.074. 11

Park, J.J., Tobet, S.A., Baum, M.J., 1998. Cell death in the sexually dimorphic dorsal preoptic area/ anterior hypothalamus of perinatal male and female ferrets. *J. Neurobiol.* 34, pp. 242-52. DOI: 10.1002/(SICI)1097-4695(19980215)34:3<242::AID-NEU4>3.0.CO;2-2. 13

Parker, K.L., Rice, D.A., Lala, D.S., Ikeda, Y., Luo, X., Wong, M., Bakke, M., Zhao, L., Frigeri, C., Hanley, N.A., Stallings, N., Schimmer, B.P., 2002. Steroidogenic factor 1: an essential mediator of endocrine development. *Recent Prog. Horm. Res.* 57, pp. 19-36. 5

Pencea, V., Bingaman, K.D., Wiegand, S.J., Luskin, M.B., 2001. Infusion of brain-derived neuro-trophic factor into the lateral ventricle of the adult rat leads to new neurons in the parenchyma of the striatum, septum, thalamus, and hypothalamus. *J. Neurosci.* 21, pp. 6706-17. 10

Pérez-Martín, M., Cifuentes, M., Grondona, J.M., López-Avalos, M.D., Gómez-Pinedo, U., García-Verdugo, J.M., Fernández-Llebrez, P., 2010. IGF-I stimulates neurogenesis in the hypothalamus of adult rats. *Eur. J. Neurosci.* 31, pp. 1533-48. DOI: 10.1111/j.1460-9568.2010.07220.x. 39

Polston, E.K., Simerly, R.B., 2006. Ontogeny of the projections from the anteroventral periven-tricular nucleus of the hypothalamus in the female rat. *J. Comp. Neurol.* 495, pp. 122-32. DOI: 10.1002/cne.20874. 9, 37

Prevot, V., Bellefontaine, N., Baroncini, M., Sharif, A., Hanchate, N.K., Parkash, J., Campagne, C., de Seranno, S., 2010. Gonadotrophin-releasing hormone nerve terminals, tanycytes and neurohaemal junction remodelling in the adult median eminence: functional consequences for reproduction and dynamic role of vascular endothelial cells. *J. Neuroendocrinol.* 22, pp. 639-49. DOI: 10.1111/j.1365-2826.2010.02033.x. 11

Puelles, L, Amat, JA, Martinez-de-la-Torre, M, 1987. Segment-related, mosaic neurogenetic pattern in the forebrain and mesencephalon of early chick embryos: I. Topography of AChE-positive neuroblasts up to stage HH18. *J. Comp. Neurol.* 266, pp. 247-68. DOI: 10.1002/cne.902660210. 4

Puelles, L., Rubenstein, J.L., 1993. Expression patterns of homeobox and other putative regulatory genes in the embryonic mouse forebrain suggest a neuromeric organization. *Trends Neurosci.* 16, pp. 472-9. DOI: 10.1016/0166-2236(93)90080-6. 3, 4

Puelles, L., 2009. Contributions to neuroembryology of Santiago y Cajal (1852-1934) and Jorge F. Tello (1880-1958). *Int. J. Dev. Biol.* 53, pp. 1145-60. DOI: 10.1387/ijdb.082589lp. 10

Rakic, P., 1972. Mode of cell migration to the superficial layers of fetal monkey neocortex. *J. Comp. Neurol.* 145, pp. 61-83. DOI: 10.1002/cne.901450105. 10, 21

Rakic, P. 1990. Principles of neural cell migration. *Experientia.* 46, pp. 882-91. 21, 25

Rakic, P., 2006. A century of progress in corticoneurogenesis: from silver impregnation to genetic engineering. *Cereb. Cortex 16 Suppl 1*, pp. i3-17. DOI: 10.1093/cercor/bhk036. 10

Rash, B.G., Grove E.A., 2011. Shh and Gli3 regulate formation of the telencephalic-diencephalic junction and suppress an isthmus-like signaling source in the forebrain. *Dev. Biol.* 359, pp. 242-50. DOI: 10.1016/j.ydbio.2011.08.026. 8

Rodríguez, E.M., Blázquez, J.L., Pastor, F.E., Peláez, B., Peña, P., Peruzzo, B., Amat, P., 2005. Hypothalamic tanycytes: a key component of brain-endocrine interaction. *Int. Rev. Cytol.* 247, pp. 89-164. DOI: 10.1016/S0074-7696(05)47003-5. 11

Rubenstein, J.L., Rakic, P., 1999. Genetic control of cortical development. *Cereb. Cortex* 9, 521-3. DOI: 10.1093/cercor/9.6.521.20

Saaltink, D.J., Håvik, B., Verissimo, C.S., Lucassen, P.J., Vreugdenhil, E., 2012. Doublecortin and doublecortin-like are expressed in overlapping and non-overlapping neuronal cell population: Implications for neurogenesis. *J. Comp. Neurol.* 520, pp. 2805-23. DOI: 10.1002/cne.23172. 11

Salic, A., Mitchison, T.J., 2008. A chemical method for fast and sensitive detection of DNA synthesis in vivo. *Proc. Natl. Acad. Sci. U. S. A.* 105, pp. 2415-20. DOI: 10.1073/pnas.0712168105. 12, 20

Saper, C.B., Loewy, A.D., Swanson, L.W., Cowan, W.M., 1976. Direct hypothalamo-autonomic connections. *Brain Res.* 117, pp. 305-12. DOI: 10.1016/0006-8993(76)90738-1. 37

Schambra, U.B., Silver, J., Lauder, J.M., 1991. An atlas of the prenatal mouse brain: gestational day 14. *Exp. Neurol.* 114, pp. 145-83. DOI: 10.1016/0014-4886(91)90034-A. 29

Schindler, S., Geyer, S., Strauß, M., Anwander, A., Hegerl, U., Turner, R., Schönknecht, P., 2012. Structural studies of the hypothalamus and its nuclei in mood disorders. *Psychiat. Res.* 201, pp. 1-9. DOI: 10.1016/j.pscychresns.2011.06.005. 1

Schwanzel-Fukuda, M., Pfaff, D.W., 1989. Origin of luteinizing hormone-releasing hormone neurons. *Nature.* 338, pp. 161-4. 10, 25

Schwarting, G.A., Raitcheva, D., Bless, E.P., Ackerman, S.L., Tobet, S., 2004. Netrin 1-mediated chemoattraction regulates the migratory pathway of LHRH neurons. *Eur. J. Neurosci.* 19, pp. 11-20. DOI: 10.1111/j.1460-9568.2004.03094.x. 26

Scordalakes, E.M., Shetty, S.J., Rissman, E.F., 2002. Roles of estrogen receptor alpha and androgen receptor in the regulation of neuronal nitric oxide synthase. *J. Comp. Neurol.* 453, pp. 336-44. DOI: 10.1002/cne.10413. 23

Segal J.P., Stallings, N.R., Lee, C.E., Zhao, L., Socci, N., Viale, A., Harris, T.M., Soares, M.B., Childs, G. Elmquist, J.K., Parker, K.L., Friedman, J.M. 2005. Use of laser-capture microdissection for the identification of marker genes for the ventromedial hypothalamic nucleus. *J. Neurosci.* 25, 4181-8. DOI: 0.1523/JNEUROSCI.0158-05.2005. 5

Seress, L., 1980. Development and structure of the radial glia in the postnatal rat brain. *Anat. Embryol.* (Berl) 160, pp. 213-26. 21

Shimada, M., Nakamura, T., 1973. Time of neuron origin in mouse hypothalamic nuclei. *Exp. Neurol.* 41, pp. DOI: 10.1016/0014-4886(73)90187-8. 163-73. 7

Shimogori, T., Lee, D.A., Miranda-Angulo, A., Yang, Y., Wang, H., Jiang, L., Yoshida, A.C., Kataoka, A., Mashiko, H., Avetisyan, M., Qi, L., Qian, J., Blackshaw, S., 2010. A genomic atlas of mouse hypothalamic development. *Nat. Neurosci.* DOI: 10.1038/nn.2545. 13, pp. 767-75. 8

Silver, J., Edwards, M.A., Levitt, P., 1993. Immunocytochemical demonstration of early appearing astroglial structures that form boundaries and pathways along axon tracts in the fetal brain. *J. Comp. Neurol.* 328, pp. 415-36. DOI: 10.1002/cne.903280308. 25

Simonian, S.X., Herbison, A.E. 2001. Differing, spatially restricted roles of ionotropic glutamate receptors in regulating the migration of gnrh neurons during embryogenesis. *J. Neurosci.*, 21, pp. 934-43. 25

Stallings, N.R., Hanley, N.A., Majdic, G., Zhao, L., Bakke, M., Parker, K.L., 2002. Development of a transgenic green fluorescent protein lineage marker for steroidogenic factor 1. *Endocr Res.* 28, pp. 497-504. 20

Steindler, D.A., Cooper, N.G., Faissner, A., Schachner, M., 1989. Boundaries defined by adhesion molecules during development of the cerebral cortex: the J1/tenascin glycoprotein in the mouse somatosensory cortical barrel field. *Dev. Biol.* 131, pp. 243-60. DOI: 10.1016/S0012-1606(89)80056-9. 25

Sugiyama, N., Kanba, S., Arita, J., 2003. Temporal changes in the expression of brain-derived neurotrophic factor mRNA in the ventromedial nucleus of the hypothalamus of the developing rat brain. *Brain Res. Mol. Brain Res.* 115, pp. 69-77. DOI: 10.1016/S0169-328X(03)00184-0. 10

Swaab D.F., 2004. Neuropeptides in hypothalamic neuronal disorders. *Int. Rev. Cytol.*, 240, pp. 305-75. DOI: 10.1016/S0074-7696(04)40003-5. 1

Swanson, L.W., Cowan, W.M., 1979. The connections of the septal region in the rat. *J. Comp. Neurol.* 186, pp. 621-655. DOI: 10.1002/cne.901860408. 6

Szarek, E., Cheah, P.S., Schwartz, J., Thomas, P., 2010. Molecular genetics of the developing neuroendocrine hypothalamus. *Mol. Cell. Endocrinol.* 323, pp. 115-23. DOI: 10.1016/j.mce.2010.04.002. 4, 9

Ter Horst, G.J., Luiten, P.G., 1986. The projections of the dorsomedial hypothalamic nucleus in the rat. *Brain Res. Bull.* 16, pp. 231-48. DOI: 10.1016/0361-9230(86)90038-9. 35

Tobet, S.A., Fox, T.O., 1989. Sex- and hormone-dependent antigen immunoreactivity in developing rat hypothalamus. *Proc. Natl. Acad. Sci. U. S. A.* 86, pp. 382-6. 19, 23

Tobet, S.A., Whorf, R.C., Schwarting, G.A., Fischer, I., Fox TO., 1991. Differential hormonal modulation of brain antigens recognized by the AB-2 monoclonal antibody. *Dev. Brain Res.* 62, pp. 91-98. DOI: 10.1016/0165-3806(91)90193-M. 23

Tobet, S.A., Fox, T.O., 1992. Sex differences in neural morphology influenced hormonally throughout life. In *Sexual Differentiation: A Lifespan Approach.* Eds, A.A. Gerall, H. Moltz and I.L., Ward. *Handbook of Behavioral Neurobiology.* 11, pp. 41-83. 6

Tobet, S.A., Chickering, T.W., Fox, T.O., Baum, M.J., 1993. Sex and regional differences in intracellular localization of estrogen receptor immunoreactivity in adult ferret forebrain. *Neuroendocrinology.* 58, pp. 316-24. DOI: 10.1159/000126556

Tobet, S.A., Crandall, J.E., Schwarting, G.A., 1993. Relationship of migrating luteinizing hormone-releasing hormone neurons to unique olfactory system glycoconjugates in embryonic rats. *Dev. Biol.* 155, pp. 471-82. DOI: 10.1006/dbio.1993.1045. 25, 33

Tobet, S.A., Chickering, T.W., Hanna, I., Crandall, J.E., Schwarting, G.A., 1994. Can gonadal steroids influence cell position in the developing brain? *Horm. Behav.* 28, pp. 320-327. DOI: 10.1006/hbeh.1994.1028. 23

Tobet, S.A., Paredes, R.G., Chickering, T.W., Baum, M.J., 1995. Telencephalic and diencephalic origin of radial glial processes in the developing preoptic area/anterior hypothalamus. *J. Neurobiol.* 26, pp. 75-86. DOI: 10.1002/neu.480260107. 5, 10, 22

Tobet, S.A., Henderson, R.G., Whiting, P.J., Sieghart, W., 1999. Special relationship of gamma-aminobutyric acid to the ventromedial nucleus of the hypothalamus during embryonic development. *J. Comp. Neurol.* 405, pp. 88-98. DOI: 10.1002/(SICI)1096-9861(19990301)405:1<88::AID-CNE7>3.0.CO;2-0. 29

Tobet, S.A., Walker, H.J., Seney, M.L., Yu, K.W., 2003. Viewing cell movements in the developing neuroendocrine brain. *Integr. Comp. Biol.* 43, pp. 794-801. DOI: 10.1093/icb/43.6.794. 20

Tobet, S.A., Schwarting, G.A., 2006. Minireview: recent progress in gonadotropin-releasing hormone neuronal migration. *Endocrinology,* 147, pp. 1159-65. 19

Tobet, S., Knoll, J.G., Hartshorn, C., Aurand, E., Stratton, M., Kumar, P., Searcy, B., McClellan, K., 2009. Brain sex differences and hormone influences: a moving experience? *J. Neuroendocrinol.* 21, pp. 387-92. DOI: 10.1111/j.1365-2826.2009.01834.x. 9

Tran, P.V., Lee, M.B., Marin, O., Xu, B., Jones, K.R., Reichardt, L.F., Rubenstein, J.R., Ingraham, H.A., 2003. Requirement of the orphan nuclear receptor SF-1 in terminal differentiation of ventromedial hypothalamic neurons. *Mol. Cell Neurosci.* 22, pp. 441-53. DOI: 10.1016/S1044-7431(03)00027-7. 7

Tsai, P.S., Brooks, L.R., Rochester, J.R., Kavanaugh, S.I., Chung, W.C., 2011. Fibroblast growth factor signaling in the developing neuroendocrine hypothalamus. *Front. Neuroendocrinol.* 32, pp. 95-107. DOI: 10.1016/j.yfrne.2010.11.002

Tsukahara, S., 2009. Sex differences and the roles of sex steroids in apoptosis of sexually dimorphic nuclei of the preoptic area in postnatal rats. *J. Neuroendocrinol.* 21, pp. 370-6. DOI: 10.1111/j.1365-2826.2009.01855.x. 9, 13

Valiente, M., Marín, O., 2010. Neuronal migration mechanisms in development and disease. *Curr. Opin. Neurobiol.*, 20, pp. 68-78. DOI: 10.1016/j.conb.2009.12.003

Waters, E.M., Simerly, R.B., 2009. Estrogen induces caspase-dependent cell death during hypothalamic development. *J. Neurosci.* 29, pp. 9714-8. DOI: 10.1523/JNEUROSCI.0135-09.2009. 13

Wierman, M.E., Kiseljak-Vassiliades, K., Tobet, S., 2011. Gonadotropin-releasing hormone (GnRH) neuron migration: initiation, maintenance and cessation as critical steps to ensure normal reproductive function. *Front. Neuroendocrinol.* 32, pp. 43-52. DOI: 10.1016/j.yfrne.2010.07.005. 26

Williams, S., Leventhal, C., Lemmon, V., Nedergaard, M., Goldman, S.A., 1999. Estrogen promotes the initial migration and inception of NgCAM-dependent calcium-signaling by new neurons of the adult songbird brain. *Mol. Cell Neurosci.* 13, pp. 41-55. DOI: 10.1006/mcne.1998.0729. 23

Wolfe, C.A., Van Doren, M., Walker, H.J., Seney, M.L., McClellan, K.M., Tobet, S.A., 2005. Sex differences in the location of immunochemically defined cell populations in the mouse preoptic area/anterior hypothalamus. *Dev. Brain Res.* 157, pp. 34-41. DOI: 10.1016/j.devbrainres.2005.03.001. 23

Wray, S., Nieburgs, A., Elkabes, S., 1989. Spatiotemporal cell expression of luteinizing hormone-releasing hormone in the prenatal mouse: evidence for an embryonic origin in the olfactory placode. *Brain Res. Dev. Brain Res.* 46, pp. 309-18. 10, 25

Xu, Y., Tamamaki, N., Noda, T., Kimura, K., Itokazu, Y., Matsumoto, N., Dezawa, M., Ide, C., 2005. Neurogenesis in the ependymal layer of the adult rat 3rd ventricle. *Exp. Neurol.* 192, pp. 251-64. DOI: 10.1016/j.expneurol.2004.12.021. 11

Xu, C., Fan, C.M., 2007. Allocation of paraventricular and supraoptic neurons requires Sim1 function: a role for a Sim1 downstream gene PlexinC1. *Mol. Endocrinol.* 21, pp. 1234-45. DOI: 10.1210/me.2007-0034

Xu, Y., Nedungadi, T.P., Zhu, L., Sobhani, N., Irani, B.G., Davis, K.E., Zhang, X., Zou, F., Gent, L.M., Hahner, L.D., Khan, S.A., Elias, C.F., Elmquist, J.K., Clegg, D.J., 2011. Distinct hypothalamic neurons mediate estrogenic effects on energy homeostasis and reproduction. *Cell Metab.* 14, pp. 453-65. DOI: 10.1016/j.cmet.2011.08.009. 33

Zeng, C., Pan, F., Jones, L.A., Lim, M.M., Griffin, E.A., Sheline, Y.I., Mintun, M.A., Holtzman, D.M., Mach, R.H., 2010. Evaluation of 5-ethynyl-2'-deoxyuridine staining as a sensitive and reliable method for studying cell proliferation in the adult nervous system. *Brain Res.* 1319, pp. 21-32. DOI: 10.1016/j.brainres.2009.12.092. 20

Zhao, L., Bakke, M., Krimkevich, Y., Cushman, L.J., Parlow, A.F., Camper, S.A., Parker, K.L., 2001. Steroidogenic factor 1 (SF1) is essential for pituitary gonadotrope function. *Development.* 128, pp. 147-54. 33

Zuloaga, D.G., Carbone, D.L., Quihuis, A., Hiroi, R., Chong, D.L., Handa, R.J., 2012. Perinatal dexamethasone-induced alterations in apoptosis within the hippocampus and paraventricular nucleus of the hypothalamus are influenced by age and sex. *J. Neurosci. Res.* 90, pp. 1403-12. DOI: DOI: 10.1002/jnr.23026. 13

Author Biographies

Stuart Tobet, Ph.D., is Professor of Biomedical Sciences and Biomedical Engineering at the Colorado State University. He received his B.S. degree in Physiological Psychology from Tulane University, New Orleans, Louisiana in 1978 and his Ph.D. in Neural and Endocrine Regulation from the Massachusetts Institute of Technology in Cambridge, Massachusetts in 1985.

His interest in brain development originated from his undergraduate experiences and research projects in the laboratory of Janis Dunlap, Ph.D. and Arnold Gerall, Ph.D. that involved looking at the impact of factors during pregnancy on long-term consequences for physiology and behavior at Tulane University in New Orleans.

As a graduate student at the Massachusetts Institute of Technology, he worked with Michael Baum, Ph.D., to identify and begin to characterize the first sexual dimorphism in the hypothalamus of a non-rodent species, ferrets.

In 1985, he joined the laboratory of Thomas Fox, Ph.D. at Harvard Medical School as a postdoctoral fellow to begin defining biochemical and cellular mechanisms that might drive the development of sexual dimorphisms in the hypothalamus.

In 1989, he joined the faculty in the Department of Neurology at Harvard Medical School with a laboratory at the E.K. Shriver Center for Mental Retardation in Waltham, Massachusetts. The laboratory goals remained to define biochemical and cellular mechanisms that drive the development of sexual dimorphisms in the hypothalamus.

In 2000, the University of Massachusetts Medical School acquired the E.K. Shriver Center and he became Associate Professor of Physiology. Together with Gerald Schwarting, Ph.D., they were the first to visualize the movement of neurons containing gonadotropin-releasing hormone from the nasal compartment into the brain using a head slice model in vitro.

In 2003, he moved to join the faculty in the Department of Biomedical Sciences in the College of Veterinary Medicine and Biomedical Sciences at Colorado State University. In 2006, he began a series of collaborations starting first with one, then 2, then 3 and more members of what became the School of Biomedical Engineering in 2007.

In 2010, he became the director of the School of Biomedical Engineering with the goals of promoting collaboration and working to solve unmet medical needs. His current research interests still revolve around determining biochemical and cellular mechanisms that drive development. However, currently "development" has expanded beyond the hypothalamus to include other brain regions, brain angiogenesis, other organs, and potential implications in cancer. Biomedical engineering projects revolve around measuring critical molecules using novel biosensors that account for high spatial and temporal resolution. Such technologies make it possible to consider assessing molecular gradients that provide key signals for cellular communications.

Dr. Tobet has received a number of grants from NSF and NIH and he is just finishing a four-year term as a standing member of the NIH Neuroendocrinology, Neuroimmunology, Rhythms, and Sleep review group, in which he served his last two years as the chairperson. He is an Associate Editor of the *Journal of Neuroendocrinology*, and serves on the editorial board of *International Journal of Endocrinology, Frontiers in Neuroendocrine Science*, the Faculty of 1000, and a new journal entitled *Technology Transfer & Entrepreneurship* to be published by Bentham Science Publishers.

Kristy McClellan, Ph.D., is an Assistant Professor of Biology at Buena Vista University. She received her B.A. degree in Biology from Taylor University in 2001 and her Ph.D. in Cell and Molecular Biology from Colorado State University in 2008. Her interest in brain development began as a first-year graduate student at Colorado State University where she was influenced by work in the labs of Robert Handa, Ph.D., and Stuart Tobet, Ph.D., that involved looking at the effects of hormones on brain development. She went on to complete her thesis work in the lab of Stuart Tobet, Ph.D., studying the effects of GABA on hypothalamic development. Currently, she is an Assistant Professor of Biology at an undergraduate university in northwest Iowa where she teaches and oversees a variety of undergraduate research projects investigating brain development.

SERIES OF RELATED INTEREST

Colloquium Series on the Building Blocks of the Cell: Cell Structure and Function

Editor

Ivan Robert Nabi, *Professor, University of British Columbia, Department of Cellular and Physiological Sciences*

This Series is a comprehensive, in-depth review of the key elements of cell biology including 14 different categories, such as Organelles, Signaling, and Adhesion. All important elements and interactions of the cell will be covered, giving the reader a comprehensive, accessible, authoritative overview of cell biology. All authors are internationally renowned experts in their area.

For a full list of published and forthcoming titles:
http://www.morganclaypool.com/page/bbc

Colloquium Series on
The Cell Biology of Medicine

Editors

Philip L. Leopold, *Ph.D., Professor and Director, Department of Chemistry, Chemical Biology, & Biomedical Engineering, Stevens Institute of Technology*

Joel Pardee, *Ph.D. President, Neural Essence; formerly Associate Professor and Dean of Graduate Research, Weill Cornell School of Medicine*

In order to learn we must be able to remember, and in the world of science and medicine we remember what we envision, not what we hear. It is with this essential precept in mind that we offer the Cell Biology of Medicine series. Each book is written by faculty accomplished in teaching the scientific basis of disease to both graduate and medical students. In this modern age it has become abundantly clear that everyone is vastly interested in how our bodies work and what has gone wrong in disease. It is likewise evident that the only way to understand medicine is to engrave in our mind's eye a clear vision of the biological processes that give us the gift of life. In these lectures, we are dedicated to holding up for the viewer an insight into the biology behind the body. Each lecture demonstrates cell, tissue and organ function in health and disease. And it does so in a visually striking style. Left to its own devices, the mind will quite naturally remember the pictures. Enjoy the show.

For a list of published and forthcoming titles:

http://www.morganclaypool.com/toc/cbm/1/1

Colloquium Series on Developmental Biology

Editors

Jean-Pierre Saint-Jeannet, *Ph.D., Professor, Department of Basic Science & Craniofacial Biology, College of Dentistry, New York University*

Daniel S. Kessler, *Ph.D., Associate Professor of Cell and Developmental Biology, Chair, Developmental, Stem Cell and Regenerative Biology Program of CAMB, University of Pennsylvania School of Medicine*

Developmental biology is in a period of extraordinary discovery and research. This field will have a broad impact on the biomedical sciences in the coming decades. Developmental Biology is interdisciplinary and involves the application of techniques and concepts from genetics, molecular biology, biochemistry, cell biology, and embryology to attack and understand complex developmental mechanisms in plants and animals, from fertilization to aging. Many of the same genes that regulate developmental processes underlie human regulatory gene disorders such as cancer and serve as the genetic basis of common human birth defects. An understanding of fundamental mechanisms of development is providing a basis for the design of gene and cellular therapies for the treatment of many human diseases. Of particular interest is the identification and study of stem cell populations, both natural and induced, which is opening new avenues of research in development, disease, and regenerative medicine. This eBook series is dedicated to providing mechanistic and conceptual insight into the broad field of Developmental Biology. Each eBook is intended to be of value to students, scientists and clinicians in the biomedical sciences.

For a full list of published and forthcoming titles:
http://www.morganclaypool.com/toc/deb/1/1

Colloquium Series on
The Genetic Basis of Human Disease

Editor

Michael Dean, *Ph.D., Head, Human Genetics Section, Senior Investigator, Laboratory of Experimental Immunology National Cancer Institute (at Frederick)*

This series will explore the genetic basis of human disease, documenting the molecular basis for rare and common Mendelian and complex conditions. The series will overview the fundamental principles in understanding such as Mendel's laws of inheritance, and genetic mapping through modern examples. In addition current methods (GWAS, genome sequencing) and hot topics (epigenetics, imprinting) will be introduced through examples of specific diseases.

For a full list of published and forthcoming titles:
`http://www.morganclaypool.com/page/gbhd`

Colloquium Series on
Genomic and Molecular Medicine

Editor

Professor Dhavendra Kumar, *MD, FRCP, FRCPCH, FACMG, Consultant in Clinical Genetics, All Wales Medical Genetics Service Genomic Policy Unit, The University of Glamorgan, UK Institute of Medical Genetics, Cardiff University School of Medicine, University Hospital of Wales*

From 1970 onwards, there has been a continuous and growing recognition of the molecular basis of medical practice. Alongside the developments and progress in molecular medicine, new and rapid discoveries in genetics have led to an entirely new approach to the practice of clinical medicine. Until recently the field of genetic medicine has largely been restricted to the diagnosis of disease, offering explanation and assistance to patients and clinicians in dealing with a number of relatively uncommon inherited disorders. However, since the completion of the human genome in 2003 and several other genomes, there is now a plethora of information available that has attracted the attention of molecular biologists and allied researchers. A new biological science of Genomics is now with us, with far reaching dimensions and applications.

During the last decade, rapid progress has been made in new genome-level diagnostic and prognostic laboratory methods, and revealing findings in genomics have led to changes in our understanding of fundamental concepts in cell and molecular biology. It may well be that evolutionary and morbid changes at the genome level could be the basis of normal human variation and disease. Applications of individual genomic information in clinical medicine have led to the prospect of robust evidence-based personalized medicine, and genomics has led to the discovery and development of a number of new drugs with far reaching implications in pharmacotherapeutics. The existence of Genomic Medicine around us is inseparable from molecular medicine, and it contains tremendous implications for the future of clinical medicine.

For a full list of published and forthcoming titles:
http://www.morganclaypool.com/toc/gmm/1/1

Colloquium Series on Integrated Systems Physiology: From Molecule to Function to Disease

Editors

D. Neil Granger, *Ph.D., Boyd Professor and Head of the Department of Molecular and Cellular Physiology at the LSU Health Sciences Center, Shreveport*

Joey P. Granger, *Ph.D., Billy S. Guyton Distinguished Professor, Professor of Physiology and Medicine, Director of the Center for Excellence in Cardiovascular-Renal Research, and Dean of the School of Graduate Studies in the Health Sciences at the University of Mississippi Medical Center*

Physiology is a scientific discipline devoted to understanding the functions of the body. It addresses function at multiple levels, including molecular, cellular, organ, and system. An appreciation of the processes that occur at each level is necessary to understand function in health and the dysfunction associated with disease. Homeostasis and integration are fundamental principles of physiology that account for the relative constancy of organ processes and bodily function even in the face of substantial environmental changes. This constancy results from integrative, cooperative interactions of chemical and electrical signaling processes within and between cells, organs and systems. This eBook series on the broad field of physiology covers the major organ systems from an integrative perspective that addresses the molecular and cellular processes that contribute to homeostasis. Material on pathophysiology is also included throughout the eBooks. The state-of the art treatises were produced by leading experts in the field of physiology. Each eBook includes stand-alone information and is intended to be of value to students, scientists, and clinicians in the biomedical sciences. Since physiological concepts are an ever-changing work-in-progress, each contributor will have the opportunity to make periodic updates of the covered material.

For a full list of published and forthcoming titles:

http://www.morganclaypool.com/toc/isp/1/1

Colloquium Series on Neurobiology of Alzheimer's Disease

Editors:

George Perry, *Ph.D., Professor of Biology and Dean of the College of Sciences, University of Texas, San Antonio*

Rudolph J. Castellani, *M.D., Professor, Pathology, University of Maryland, School of Medicine*

This e-book series on Alzheimer's disease will provide an up-to-date, comprehensive overview of Alzheimer's disease and dementia from a multidisciplinary perspective, with a focus on disease pathogenesis and translational neurobiology. The major pathogenic proteins will be described and discussed in depth from the perspective of molecular and cell biology, experimental and transgenic modeling, and in-situ phenotypic expression in humans. Added to this will be in depth discussions of all the major pathogenic theories, including the amyloid and tau protein cascades, oxidative stress, involvement of heavy metals, interplay of the endocrine system, issues surrounded cell cycle activation and protein signaling, and important comorbidities that influence human disease such as vascular neurobiology and synucleinopathy. Treatment paradigms and trials as a function of known components of disease pathogenesis will also be described and discussed in requisite detail that reflects the state of the art. As such, the eBook series will provide a "one stop shop" for the aspiring neuroscientist or physician scientist, and will prove an adaptable framework that can be updated going forward, as dictated by new discoveries and the accumulating scientific literature.

For a list of published and forthcoming titles:

http://www.morganclaypool.com/page/alz

Colloquium Series on Neuropeptides

Editors

Lakshmi Devi, *Ph.D., Professor, Department of Pharmacology and Systems Therapeutics, Associate Dean for Academic Enhancement and Mentoring, Mount Sinai School of Medicine, New York*
Lloyd D. Fricker, *Ph.D., Professor, Department of Molecular Pharmacology, Department of Neuroscience, Albert Einstein College of Medicine, New York*

Communication between cells is essential in all multicellular organisms, and even in many unicellular organisms. A variety of molecules are used for cell-cell signaling, including small molecules, proteins, and peptides. The term 'neuropeptide' refers specifically to peptides that function as neurotransmitters, and includes some peptides that also function in the endocrine system as peptide hormones. Neuropeptides represent the largest group of neurotransmitters, with hundreds of biologically active peptides and dozens of neuropeptide receptors known in mammalian systems, and many more peptides and receptors identified in invertebrate systems. In addition, a large number of peptides have been identified but not yet characterized in terms of function. The known functions of neuropeptides include a variety of physiological and behavioral processes such as feeding and body weight regulation, reproduction, anxiety, depression, pain, reward pathways, social behavior, and memory. This series will present the various neuropeptide systems and other aspects of neuropeptides (such as peptide biosynthesis), with individual volumes contributed by experts in the field.

For a list of published and forthcoming titles:

`http://www.morganclaypool.com/toc/npe/1/1`

Colloquium Series on Neuroglia: From Physiology to Disease

Editors

Alexej Verkhratsky, *Ph.D., Professor of Neurophysiology, University of Manchester*
Vlad Parpura, Ph.D., Associate Professor of Neurophysiology, University of Alabama at Birmingham

For decades the neuroglia were known, and known to be numerous, in the neural system. Nevertheless they were long thought to play only a minor, supporting role to other cells such as axons and neurons. Glial cell are now recognized are essential to neural functioning and represent an exciting, rapidly growing field in the neurosciences. This series will explore the overall molecular physiology of glial cells as well as their role in pathologic conditions.

For a full list of published and forthcoming titles:
`http://www.morganclaypool.com/page/neuroglia`

Colloquium Series on
Protein Activation and Cancer

Editor

A. Majid Khatib, *Ph.D. Research Director, INSERM, and, University of Bordeaux*

This series is designed to summarize all aspects of protein maturation by proprotein convertases in cancer. Topics included deal with the importance of these processes in the acquisition of malignant phenotypes by tumor cells, induction of tumor growth, and metastasis. This series also provides the latest knowledge on the clinical significance of convertase expression and activity, and the maturation of their protein substrates in various cancers. The potential use of their inhibition as a therapeutic approach is also explored.

For a full list of published and forthcoming titles:
`http://www.morganclaypool.com/page/pac/1/1`

Colloquium Series on Stem Cell Biology

Editor

Wenbin Deng, *Ph.D., Cell Biology and Human Anatomy, Institute for Pediatric Regenerative Medicine, School of Medicine, University of California, Davis*

This Series is interested in covering the fundamental mechanisms of stem cell pluripotency and differentiation, and strategies for translating fundamental developmental insights into the discovery of new therapies. The emphasis is on the roles and potential advantages of stem cells in developing, sustaining and restoring tissue after injury or disease. Some of the topics covered include: the signaling mechanisms of development and disease; the fundamentals of stem cell growth and differentiation; the utilities of adult (somatic) stem cells, induced pluripotent stem (iPS) cells and human embryonic stem (ES) cells for disease modeling and drug discovery; and finally, the prospects for applying the unique aspects of stem cells for regenerative medicine. We hope this Series will provide the most accessible and current discussions of the key points and concepts in the field, and that students and researchers all over the world will find these in-depth reviews to be useful.

For a list of published and forthcoming titles:
`http://www.morganclaypool.com/page/scb`